- Why do volcanoes erupt, and in so many different ways?

- What happens when a volcano erupts in the sea?

- How can eruptions be predicted?

These are but a few of the many questions addressed in this clearly written and wide-ranging introduction to volcanoes. Assuming little background knowledge, and providing a comprehensive glossary explaining technical terms, the book deals with all aspects of volcanic features and processes. It offers a fresh perspective in terms of volcanoes as distinctive and often dramatic features in the landscape, as well as potential sources of hazard to human beings.

Volcanoes provide an exciting way in which to analyze some of the major – and visually most impressive – processes on Earth. This book sets them in their world context and explains their formation, especially in relation to their many styles of eruption and the multitude of volcanic landscape forms that result. Some major eruptions are selected to illustrate this eruptive variety and the impact on the surrounding populations. Distinctive landscape forms, from flows to cones and calderas, are described with reference to the "biographies" of the volcanoes. There are chapters specifically on stratovolcanoes, hydrovolcanic features and erosional processes – all of which have not been well treated elsewhere. A final chapter examines the latest methods of predicting volcanic eruptions and moderating their effects.

Drawing on an unusually wide range of sources in the French, Spanish and Portuguese literature, as well as English, the author presents examples and illustrations from around the world, including the Aegean, the French West Indies, the American West, the Azores and the Canary Islands, Italy and central France. He has also translated various eye-witness accounts of volcanic events and has included a revised version of Pliny's account of the eruption of Vesuvius.

Fully illustrated with photographs and line drawings, *Volcanoes* is an unusually accessible introduction to a perennially fascinating subject and is sure to be read by not only students of geography, the Earth sciences and the environmental sciences, but also a wider readership by whom the book's clarity and minimal technical jargon will be warmly welcomed.

Alwyn Scarth was born and educated in the West Riding of Yorkshire. He graduated from the University of Cambridge, where he was also awarded a PhD, and was for over thirty years Lecturer in Geography at the University of Dundee; he also spent some time at the University of Clermont-Ferrand while doing research for his PhD in central France. His studies of volcanoes have extended across Europe, the Mediterranean, the Canaries, the Azores, Madeira and the United States. Among many publications, he has written on the volcanic nature of the Polyphemus tale in Homer's Odyssey.

The Louise Lindsey Merrick
Natural Environment Series

Number Nineteen

Volcanoes

An introduction

Alwyn Scarth

UCL PRESS

First published in 1994 by UCL Press

UCL Press Limited
University College London
Gower Street
London WC1E 6BT

The name of University College London (UCL) is a registered
trade mark used by UCL Press with the consent of the owner.

ISBN:
1-85728-223-X HB
1-85728-224-8 PB

British Library Cataloguing-in-Publication Data
A catalogue record for this publication is available
from the British Library.

Typeset in 10/12pt Bembo
Printed and bound by
Butler & Tanner, Frome, England.

To my nephews and nieces

Contents

Preface

I want you to read this book with pleasure and share some of the enjoyment that I have had in studying, visiting and admiring volcanoes. I should like to introduce volcanoes in their fascinating variety to those who wish to learn more about what are, I think, the most beautiful landscapes in the world. The rich but often complicated and dispersed expert literature on volcanoes, combined with my own observations, form the basis of this book and I have tried to include many of the latest ideas in a rapidly advancing subject. However, direct references that would have made the text too heavy have been eliminated, although, for those whose appetite has been whetted, a selection of further reading is indicated at the end. I have also tried to avoid jargon and I use a very minimum of technical terms, but I hope that the glossary will provide a useful means of clarification where they could not be avoided.

I must stress that, as a geographer, I am most interested in the geographical aspects of volcanoes, although, as several chapters in this book show, they are inseparable from their geological origins. Most books on volcanoes emphasize either the generation of volcanic materials beneath the Earth's surface, or the geological variety of the products that they expel. I have, instead, concentrated on aspects of volcanoes as landscape features and the different impacts that their eruptions have had upon the landscape and populations around them. The examples chosen were drawn wherever possible from areas that I know best – some will therefore seem to betray a regional bias, but many will also perhaps be unfamiliar, although, I hope, none the less appropriate. In the end, if reading this book encourages you to visit volcanoes and it enables you to see them with new eyes, then my own efforts and yours will have been worthwhile. But perhaps you should read the chapter on volcanic prediction before you go, just to avoid taking unnecessary risks.

Among the many friends who in their various ways have offered stimulus and encouragement, I should especially like to express my appreciation to my colleague Rob Watt. I should also like to thank the Trustees of the Carnegie Fund for the Universities of Scotland for a generous grant that enabled me to study in the Azores. I am grateful to Dr P. Cattermole for valuable instruction on the generation of magma, and to Dr J. C. Carracedo for greatly improving my knowledge of the eruption of Lanzarote. I thank John Brush, Juan Carlos Carracedo, Jean-Paul Gadet, Raymundo Punongbayan, Josette Tourenq and Margot Watt for providing me with photographs for some of the illustrations, Jim Ford for drawing all the diagrams, and Pat Michie for secretarial help.

Alwyn Scarth   Dundee and Paris
                                                                             1993

Acknowledgements

The author and publishers would like to thank the following for permission to make use of copyright material in this book:

Springer Verlag and:

H. Sigurdsson et al., "The 1982 eruptions of El Chichón volcano, Mexico", *Bulletin Volcanologique* **49**(2) (1987), fig. 2, p. 470 [for Fig. 6.3].

E. Tryggvason, "Widening of the Krafla fissure swarm during the 1975–81 volcano-tectonic episode", *Bulletin Volcanologique* **47** (1984), fig. 2, p. 50 [for Fig. 15.4].

F. Barberi et al., "Phlegraean fields 1982–1984: brief chronicle of a volcano emergency in a densely populated area", *Bulletin Volcanologique* **47** (1984), fig. 2, p. 177; fig. 3, p. 178 [for Figs 15.2, 15.3].

R. B. Moore, "Volcanic geology and eruption frequency, São Miguel, Azores", *Bulletin Volcanologique* **52** (1990), fig. 3, p. 604 [for Fig. 11.3].

P. Imsland, in *Volcanic hazards assessment and monitoring*, J. H. Latter (ed.) (1989), fig. 2, p. 40 [for Fig. 5.4].

Elsevier Science Publishers BV and:

C. R. Bacon, "Eruptive history of Mount Mazama and Crater Lake Caldera, Cascade Range, USA", *Journal of Volcanology and Geothermal Research* **18** (1983), pp. 17–115, fig. 1, p. 61 [for Fig. 11.5].

B. Voight, "The 1985 Nevado del Ruiz volcano catastrophe: anatomy and retrospection", *Journal of Volcanology and Geothermal Research* **42** (1990), pp. 151–88, quotation from p. 173.

I. Yokoyama et al., "Energy partition in the 1982 eruption of El Chichón Volcano, Chiapas, Mexico", *Journal of Volcanology and Geothermal Research* **51** (1992), fig. 1, p. 2 [for Fig. 6.3].

F. LeGuern et al., "Witness accounts of the catastrophic event of August 1986 at Lake Nyos (Cameroon)", *Journal of Volcanology and Geothermal Research* **51** (1992), fig. 1, p. 172 [for Fig. 5.13].

F. Barberi et al., "The control of lava flow during the 1991–1992 eruption of Mount Etna", *Journal of Volcanology and Geothermal Research* **56** (1993), fig. 1, p. 3 [for Fig. 15.6].

R. S. J. Sparks, "The dynamics of bubble formation and the growth of magmas: a review and analysis", *Journal of Volcanology and Geothermal Research* **3** (1978), fig. 14, p. 31 [for Fig. 4.2].

C. Deniel et al., "^{230}Th ^{238}U and ^{14}C age determinations from Piton des Neiges Volcano Réunion", *Journal of Volcanology and Geothermal Research* **51** (1992), fig. 1, p. 254 [for Fig. 5.7].

R. A. Duncan et al., "Réunion hotspot activity through Tertiary time", *Journal of Volcanology and Geothermal Research* **36** (1989), fig. 2, p. 196 [for Fig. 5.6].

J. C. Carracedo et al., "The 1730–1736 eruption of Lanzarote, Canary Islands: a long high magnitude basaltic fissure eruption", *Journal of Volcanology and Geothermal Research*, vol.53 (1992), fig.1A, p. 240 [for Fig. 6.29].

H. Sigurdsson et al., "Pre-eruption compositional gradients and mixing of andesite and dacite magma erupted from Nevado del Ruiz volcano, Colombia in 1985", *Journal of Volcanology and Geothermal Research* **41** (1990), fig. 1, p. 128 [for Fig. 6.4].

Acknowledgements

P. Mitropoulos et al., "Petrogenesis of Cenozoic volcanic rocks from the Aegean Island Arc", *Journal of Volcanology and Geothermal Research* **32** (1987), fig. 1, p. 178 [for Fig. 6.13].

The Royal Geographical Society for A. Scarth, "The Montagne de la Serre", *Geographical Journal* **133**(1) (1967), fig. 4, p. 45 [for Fig. 14.11].

A. De Goer de Hervé et al., *Volcanologie de la Chaîne des Puys*. Parc Naturel Régional des Volcans d'Auvergne, 63 Aydat, France, figures on pp. 31, 37, 49. [for Figs 12.1 (top), 12.8, 12.7, respectively].

Thera Foundation, London for D. M. Pyle, "New estimates for the volume of the Minoan eruption', in *Thera and the Aegean World* III 1990, The Thera Foundation, London, fig. 1, p. 115 [for Fig. 6.13, inset].

Thera Foundation, London for T. H. Druitt & V. Francaviglia, "An ancient caldera cliff line at Phira and its significance for the topography and geology of pre-Minoan Santorini', *in Thera and the Aegean World III* 1990, The Thera Foundation, London, fig.4, p. 367 [for Fig. 6.14].

K. H. Wohletz & M. F. Sheridan, "Hydrovolcanic explosions II", *American Journal of Science*, **283** (1983), fig.1, p. 386; fig.2, p. 389. By permission of *American Journal of Science* [for Fig. 12.1, middle and lower].

F. Machado et al., "Capelinhos eruption of Fayal volcano, Azores 1957-1958", *Journal of Geophysical Research* **67**(9), August 1962, fig.4, p. 3524. Copyright by the American Geophysical Union [for Fig. 6.19].

P. Francis, *Volcanoes: a planetary perspective*, Formation of levees in a typical lava, fig.7.6, p. 143. Oxford University Press (1993). By permission of Oxford University Press [for Fig. 7.6, lower].

H. Williams, "Stages in the formation of a caldera", in *Calderas and their origin*, University of California Publications in Geology 26 (1941), fig. 37, p. 336 [for Fig. 11.2].

Iceland Glaciological Society and *Jökull* for K. Saemundson, "Tectonic map of Iceland", in *Jökull* **29** (1979) (*Outline of the geology of Iceland*), fig. 1, p. 8 [for Fig. 5.5].

Geological Society of London for J. G. Jones, "Intraglacial volcanoes of the Laugarvatn region, SW Iceland", *Geological Society of London, Quarterly Journal* **124** (1969), fig. 2, pp. 202–3 [for Fig. 5.10].

United States Geological Survey for:

R. S. Williams & J. G. Moore, *Man against volcano: the eruption of Heimaey, Vestmannaeyar, Iceland*, USGS Publication 1983-381-618/103 (1983), map on p. 6 [for Fig. 6.21].

Generalization of Plate 1 Map in "The 1980 eruptions of Mount St Helens, Washington', USGS Professional Paper 1250, R. W. Lipman & D. R. Mullineaux (eds) [for Fig. 6.7].

E. W. Wolfe, "The 1991 eruptions of Mount Pinatubo, Philippines", *Earthquakes and Volcanoes* **23**(1) (1992), fig. 1, p. 5 [for Fig. 6.23]; fig. 10, p. 11 [for Fig. 6.25]; fig. 18, p. 22 [for Fig. 6.26]; fig. 23, p. 27 [for Fig. 6.27].

Prentice Hall Inc. for G. A. Macdonald, *Volcanoes* (1972), fig. 8.2, p. 149. [for Fig. 5.11]. Reprinted by permission of Prentice-Hall Inc., Englewood Cliffs, New Jersey 07632, USA.

My own translations of:

− Pliny's account of the eruption of Vesuvius in AD79 used as its basis Letters 16 and 20 (To Tacitus) from *Pliny the Younger: Letters and Panegyricus*, translated by Betty Radice. Used with permission of Harvard University Press and the Trustees of the Loeb Classical Library. Copyright 1969.

−Dr Guérin's account of the lahar from Montagne Pelée is based on the original in Jean Hess: *La catastrophe de la Martinique*, Charpentier, Paris 1902 [in *Saint Pierre* (Martinique) Tome II

Acknowledgements

(*La catastrophe et ses suites*) by Solange Contour (Editions Caribéennes, Paris, 1989)].
– excerpts of the correspondence between the Royal Court of Justice and the Volcano Emergency Committee in Lanzarote (1730–1731) in the Archivo de Simancas (Gracia y Justicia, Legajo 89) were made, with permission from the Director, the Archivo General de Simancas (Spain) and J. C. Carracedo & E. R. Badiola: *Lanzarote, La erupcion Volcanica de 1730*, CSIC, Servicio de Publicaciones, Excmo. Cabildo Insular de Lanzarote (1991).
– the extracts from the diary of the Parish Priest at Yaiza, Lanzarote were from *Description physique des Isles Canaries* by L. Von Buch, translated into French by C. Boulanger, Levrault, Paris (1836).

PART I

Introducing volcanoes

1

Introduction

Volcanoes are exciting. To witness the spectacle of a great volcanic eruption is a thrilling, although dangerous, privilege. Violent eruptions can outstrip all other natural processes, releasing more energy than thunderstorms or hurricanes, and, at the same time, giving us the macabre sensations of catastrophe and disaster. But, these awesome eruptions also reward our aesthetic senses by creating volcanoes that are amongst the most graceful and beautiful of all the world's landscapes.

Although many volcanoes may seem ancient in comparison with our short lives, they are formed by some of the fastest of all the processes making the Earth's relief features, so that even the oldest active volcanoes are youthful, and their eruptions are mere flashes when seen in the perspective of geological time that stretches back 4600 million years.

Some volcanic eruptions hold the stage for only a brief moment. Monte Nuovo began to erupt in Campania near the Bay of Naples on 29 September 1538, and in 2 days a cone 130m high had been formed. But this modest eruption ceased after a week and has never resumed. On the other hand, great volcanoes such as Etna in Sicily, Popocatépetl in Mexico or Fuji in Japan have been built up by many eruptions over hundreds of thousands of years. Every phase of activity, though, has usually only been brief and interspersed by much longer periods of repose. The shortest episodes of all – such as those that decapitated Vesuvius in AD 79 and disintegrated Krakatau in 1883 – are often the most powerful climactic spasms that last only a few hours during great eruptions that themselves rarely continue for more than a couple of extremely turbulent days

The sudden violence of volcanic eruptions causes catastrophe and devastation. Until modern prediction and surveillance techniques were developed, most eruptions came as a total surprise, and destruction and death almost automatically ensued. They are made more menacing, too, because volcanic soils are often fertile; farmers have thus been tempted to cultivate more and more land within the danger zones, such that economic factors have often overridden geological warnings, until accounts have been settled on the day of reckoning. Thus, the often violent Indonesian and Japanese volcanoes can be lethal because they are surrounded by large populations. The eruption of Tambora in Indonesia in 1815, for instance, is reputed to have cost almost 92000 lives. The celebrated eruption of Vesuvius buried Pompeii and preserved a vivid relic of a Roman town. Less famous yet, but potentially of equal importance, is the entombment of Akrotiri by the Bronze Age eruption of Santorini in the Aegean Sea. Great eruptions, too, indirectly generate other features such as volcanic mudflows (known as lahars) and tidal waves (tsunamis) that spread destruction far beyond the volcano itself. Lahars at Nevado del Ruiz in Colombia in 1985 and tsunamis from Krakatau in 1883, for instance, were virtually the sole causes of the many thousands of deaths resulting from these disastrous eruptions.

Volcanic eruptions also send fine ash into the stratosphere which often circulates around the world for several years. As a result, some scientists have been inclined to blame volcanoes for both short-term and long-term climatic changes. Cold winters and damp, dismal summers in temperate latitudes have been ascribed to major eruptions. The eruption of El Chichón in Mexico in 1982 may have been responsible for the abnormally cold winters that ensued in North America. Similarly the eruption of Tambora in 1815 expelled a great ash cloud that entered the stratosphere and caused the brilliant sunsets that were an added source of inspiration to J. M. W. Turner half the world away. But this cataclysm is also reputed to have caused the poor weather of 1816, "the year without a summer", in Europe and North America. Some would claim that the Ice Ages themselves were caused by concentrations of major volcanic eruptions in the past few million years. However, in spite of a few recent studies, much of the vital evidence that might prove these theories is still locked up in the polar ice-caps awaiting analysis.

Volcanic activity is far from being wholly negative. Many volcanoes erupt harmlessly and many eruptions have their beneficial aspects. Most materials erupted by volcanoes are rich in valuable nutrients, so that a layer of ash scattered in the fields stimulates yields almost as much as artificial fertilizer. Excellent crops were produced in eastern Washington State after Mount St Helens erupted in 1980. Weathered volcanic soils, such as those of Java, for instance, and the lower slopes of Etna and Vesuvius, are amongst the richest in the world, and valuable minerals such as gold, silver, mercury, diamonds, copper, lead and zinc are often associated with volcanoes and their feeder conduits. Volcanic gas, steam and hot water have been harnessed, as in Larderello in Italy and in Chaudes Aigues in Auvergne in France, to provide cheap geothermal energy and hot water, and in Iceland more than three-quarters of the population uses geothermal heat. Volcanoes have also become major tourist attractions: Hawaii has a superb national park around Kilauea; excursions to Etna in Sicily, or to Teide in Tenerife, are regular features of trips designed to distract tourists from sun-worship. Park rangers evoke the great eruption of 1980 for the thousands who flock to the "US National Monument Park" of Mount St Helens. Volcanoes, then, are not all doom and disaster. On balance, in fact, volcanoes probably give more days of celebration than days of wrath.

Volcanoes have an aesthetic quality that perhaps no other group of landscape features can surpass. Many volcanic cones have graceful and elegant forms of outstanding and awesome beauty. Fuji was immortalized by the famous views painted by Hokusai. Popocatépetl forms the beautiful backcloth to Cholula in Mexico that astonished the Spanish Conquistadors in 1519. Santorini is the most spectacular island in the Mediterranean Sea. Vesuvius graces the renowned Bay of Naples and, at the same time, adds just the smallest frisson to the famous saying "see Naples and die". Even the fuming wreck of Mount St Helens will soon regain the ethereal beauty for which it was renowned until 1980. Volcanic areas retain their quality even on the smallest scales. The precipitations at Mammoth Springs in Yellowstone Park in the United States, and at Pamukkale in Turkey, produce delicate fluted cascades, and the crystals formed around fumaroles give miniature landscapes that no jeweller could surpass. Such beauty at all scales gives volcanic areas an irresistible glamour that adds another dimension to the rewards of studying them (Fig. 1.1).

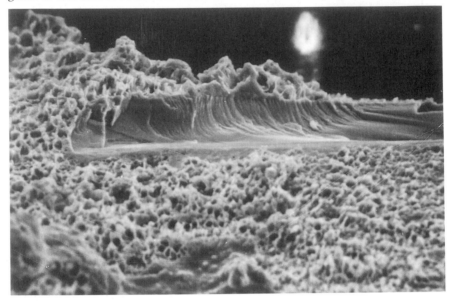

Figure 1.1 Dust erupted from El Chichón, 1982 (magnification ×500; courtesy Josette Tourenq).

A volcano is usually a cone-shaped hill or mountain composed of materials erupted through an opening in the Earth's crust which extends from the hotter zones below (Fig. 1.2). These materials, derived from gas-rich molten rock, or magma, vary from lava-flows, fine ash and cinders to steam and poisonous gases (Fig. 1.3). The eruptions are nearly always explosive. Some, however, make no more than a little popping, spluttering noise; others make great blasts that can be heard at enormous distances. All the volcanic materials are emitted from chimneys or vents pierced through the lithosphere, the solid outer shell of the Earth, whose surface layers form the crust. These vents sometimes occur singly, sometimes in closely linked clusters, or sometimes strung out along cracks or fissures. Often the magma becomes concentrated within the crust in a chamber or reservoir, where it can remain for many years before resuming its ascent, or it may transit quickly upwards by means of groups of deep fissures. The magma rises more rapidly when the pressures upon it are diminished as it approaches the Earth's surface. It becomes more bubbly and frothy as the gases within it expand, and the escape of the magma at the surface is

Figure 1.2   Stromboli by night.

Figure 1.3   Lavas covering ash, El Portillo, Tenerife, Canary Islands.

5

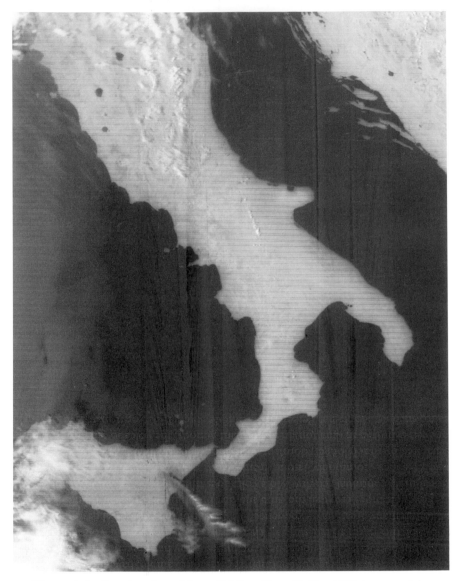

Figure 1.4 Erupting plume blown SW from Etna, 15 May 1983 (courtesy John Brush, NERC Satellite Receiving Station, University of Dundee).

sometimes likened to what happens when a bottle of warm champagne (or beer) is suddenly uncorked. When the magma contains much gas, it escapes in noisy explosions which shatter much of it into fine fragments such as ash or cinders. These fragments are hurled into the air, cool down and then fall back and often accumulate near the vent to form cones made up of layer upon layer of these bedded fragments, or are blown to high altitudes and carried away by the winds (Fig. 1.4). If the magma

Figure 1.5 Wizard cinder cone and its lava-flow, Crater Lake, Oregon.

contains little gas, some of it runs quite calmly out across the land as lava-flows (Fig. 1.5). The volcanoes usually have a bowl-shaped hollow, or crater, at the summit immediately above the vent (Fig. 1.6). The vent and the base of the crater are sealed up by solidified lava between eruptions, when the only sign of activity may be gas and steam hissing from small cavities called fumaroles.

Volcanoes are restricted to very well defined, usually fractured areas of the Earth's surface and are often found in long narrow bands. Most volcanic activity – and probably the most continuous eruptions – takes place under water, along the crests of the mid-ocean ridges, which have a total length in excess of 60000 km. Except in rare cases such as Iceland, where the ridges emerged above sea level, these eruptions were largely unknown until they were detected as a result of remarkable advances in oceanography and geophysics in recent decades. Almost two-thirds of all the presently active volcanoes on land are situated in narrow zones around the Pacific Ocean, whereas vast tracts of continental Asia, for example, have none at all. There are about 500 active land volcanoes, and just as many have recently become extinct. Most of the Earth's land surface, therefore, shows no sign of volcanic activity, although many areas have experienced eruptions in the geological past.

Few volcanoes erupt all the time and Stromboli in the Aeolian Islands is exceptional in its continual activity. Most volcanoes usually display relatively short eruptive episodes, commonly lasting from a few days to a few months, which are separated by much longer, dormant periods of repose, which may be years, decades or even centuries in duration. In its dormant periods, the volcano can be completely calm, such as Montagne Pelée in Martinique since 1932, or it may give off small

7

Figure 1.6 North-East crater on Etna in eruption. The spots visible in the air are lava fragments.

emissions of steam and gases, such as Vulcano, that betray its active state. Dormant periods often vary in length in any one volcano, so that eruptive episodes happen irregularly. Vesuvius, for example, had been dormant for several centuries before its spectacular eruptions of AD 79 and 1631, but it erupted persistently from 1872 to 1906. The irregular length and frequent calm of dormant periods mean that it is often extremely difficult to decide whether or not a volcano is extinct. If a volcano has not erupted for a million years, it is very likely to be extinct, but the problem is much more difficult if the volcano has only been calm for several millennia. The geological record may be ambiguous, or the historical record too brief or inaccurate. Indeed, in many volcanic areas, such as the Cascade Range of the United States and Canada, for instance, records do not extend back more than 200 years. Thus, some eruptions cause surprise, like those of Mount Lamington in Papua-New Guinea in 1951, and the start of that of Pinatubo in the Philippines in 1991.

On the other hand, relics of eroded volcanoes can be found from practically every period of geological history, although it is very unlikely that they will ever spring to life again. In Scotland, for example, Arthur's Seat in Edinburgh, the isolated hills or "laws" of the Midland Valley, and the Black Cuillins of Skye are all eroded stumps of ancient volcanoes. Most of the Earth's active volcanoes are much less than a million years old, many have been created in historic time and several, such as Surtsey and Parícutin, have been born this century. Others are bound to emulate them in future centuries. Volcanic eruptions and the growth of volcanoes are natural and altogether inevitable processes on our Earth.

8

2

Vulcano

Volcanoes are so named because of Vulcano. To the ancient Greeks, Vulcano was Hiera Hephaistou, sacred to Hephaistos; to the ancient Romans it was the home of the forges of Vulcan; to both, Vulcano was the lair of the god of fire, in recognition of the subterranean forces that rumbled in conflict on the island. The myths, or most of them, have vanished, but the name of Vulcano has remained a symbol for all volcanoes and has given the very word to most European languages and, indeed, to Arabic too. But it was not the only volcano known to the ancients. They also knew Stromboli and Etna, had known and forgotten Santorini, and in AD 79 Vesuvius forced itself upon their memories. Vulcano has none of the immensity of Etna, the regular daily outbursts of Stromboli, nor the great power of Vesuvius in eruption, but perhaps it was the variety of its volcanic features that impressed Vulcano upon the consciousness of the ancient world.

Vulcano is one of the Aeolian Islands north of Sicily in the central Mediterranean Sea (Fig. 2.1). It is a small island of no more than 22km^2 and it rises a mere 500m above sea level. The boat from Milazzo in Sicily offers the best approach to Vulcano, showing off the black lava in the cliffs and ravines as it hugs the east coast of the island. Soon the boat rounds a rugged lava promontory, and enters the Porto di Levante harbour (Fig. 2.2). The modern volcano, Fossa, rises to the south, and a smaller cone, Vulcanello, closes the northern shore of the bay. Fossa is only 391m high and a path guides the visitor to its bare summit. It has been dormant since it last erupted between 1888 and 1890 (Fig. 2.3).

The path climbs just below a small double crater, the Forgia Vecchia, that erupted in 1727 and now forms a large wound on the flanks of Fossa. A little farther on, the path crosses the glassy, rugged lava-flow of the Pietre Cotte, which is made of volcanic glass, obsidian. In 1739 it oozed down slope like clotting, grey porridge and solidified before it could even spread beyond the foot of the Fossa cone.

Soon the path reaches the summit of Fossa, marked by a smooth rim of fine grey ash, strewn with large, flattened, angular boulders of lava. But the main attraction here is the 175m-deep funnel-shaped crater, the Gran Cratere. Layer upon layer of fine grey, reddish, and buff ash are exposed in its walls (Fig. 2.4). It looks as if it had been formed yesterday, but there has been no eruption here since 1890. The Gran Cratere has been the site of eruptions for more than 2000 years, and Fossa itself is about 14000 years old. For example, when Virgil, in the Aeneid, poetically

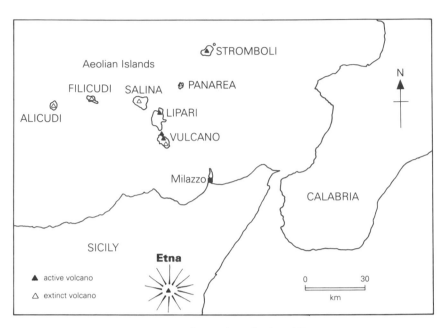

Figure 2.1   The Aeolian Islands and Etna.

described the roaring furnaces and the clang of metal on the anvil in Vulcan's forge, he was probably inspired by the noisy eruptions that took place in the Gran Cratere between 29 BC and 19 BC. At present, only the little holes of fumaroles hiss out gas, steam, and sulphurous fumes that condense to form beautiful, if very fragile, yellow crystals. These fumaroles indicate that Vulcano is dormant and only resting, and if Fossa maintains its pattern of activity, an eruption is due within a decade or so.

The view from the summit of Fossa is full of volcanic features. In the south, the older part of Vulcano ends suddenly in a circular cliff facing onto Fossa. This is the wall of the hollow, or caldera, formed when an older and larger volcano sank down before Fossa was even created. The remains of this older volcano form the southern two-thirds of the island. Northwards the view extends across the two bays, separated by a low isthmus of black sand, where the village lies. Beyond is the former island of Vulcanello that was only joined by the isthmus to Vulcano in 1550.

On the way to Vulcanello, other small volcanic features are grouped together in a beautifully situated zone on a main fault line caused by a crack in the Earth's crust stretching northwards from Fossa to Lipari Island. The Porto di Levante is guarded by the Faraglione, a decayed volcanic remnant, 56m high, altered by volcanic spring waters. Alongside it is a large pit filled with warm pale-grey mud about a metre deep, which bubbles continually. Tourists bathe in the mud and spread it all over their bodies because it is reputed to cure both rheumatism and ugliness. On the little Acqua Calda beach to the north, hot spring waters bubble from the sand and the sea in long fissure-lines. These too have curative properties, but they can also be extremely uncomfortable to sit upon.

10

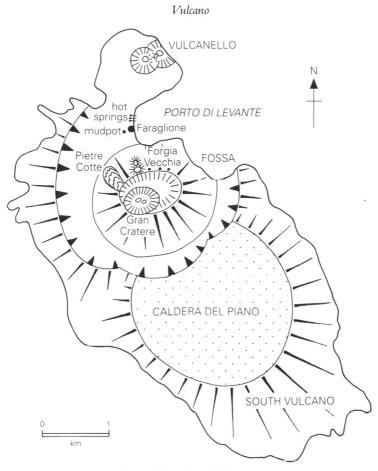

Figure 2.2 Map of Vulcano.

Vulcanello, north of the bay, is as delightful as its name suggests. It stands on a plinth of dark brown lava lying in mounds and coils as if it had been squirted from a giant tube of toothpaste. Vulcanello only rises to 123m but has three steep, fresh-looking, buff-coloured craters on its summit. It was born in Roman times in 183 BC as a submarine volcano, so it is even younger than Fossa. Its last eruption took place in 1550 and it looks as if it might resume activity at any moment. The view from Vulcanello, especially on a summer evening is very fine (Fig. 2.3). The squat, gully-scarred cone of Fossa, backed by its semi-circular protective caldera wall, forms a superb backcloth to the white cubic houses in the village and the little harbour, the hot beach, and the black sandy apron of the isthmus. Odysseus probably came to Vulcano after blinding Polyphemus at Etna. The geological record in these rocks and structures reveals great and complicated events. The historical record of volcanic action, here stretching back 2000 years, makes Vulcano still a worthy site for the home of a god. Its features demonstrate how vivid and varied volcanic activity can be, and underline that it is not always limited to lava-flows or vast death tolls.

11

Figure 2.3   Vulcano: Fossa from Vulcanello.

Figure 2.4   Vulcano: Gran Cratere on Fossa cone.

PART II

Volcanic activity

3

Distribution of volcanoes

Volcanoes do not erupt everywhere. They are restricted to narrow bands in very specific locations (Fig. 3.1). The Pacific Ocean is virtually surrounded by a belt of volcanoes commonly called the "Pacific Ring of Fire" which contains about two-thirds of the world's active volcanoes. The "Ring of Fire" is marked by the volcanic chains of Japan, Kamchatka, South Alaska and the Aleutian Islands, the Cascade Range of the United States and Canada, Central America, the Andes, New Zealand, Tonga, Vanuatu, Papua-New Guinea, Indonesia, the Philippines, and finally the Mariana, Izu and Bonin Islands, which complete the circle.

In contrast, the Atlantic Ocean has few volcanoes around its edges, except those of the Canary Islands, the Cape Verde Islands, and the West Indies. On the other hand, the Atlantic Ocean has many volcanoes along its central ridge, which are largely submerged. Its crest, however, emerges in Jan Mayen and Iceland, and volcanoes have arisen on the ridge flanks in Ascension Island, Tristan da Cunha, and in the Azores. Similarly the mid-ocean ridges of the Pacific and Indian Oceans are also marked by innumerable submerged volcanoes.

The Mediterranean Sea, too, has volcanoes in its midst in southern Italy and Greece. Other volcanoes occur in, or alongside, some of the major rift valleys: the East African Rift Valley system and the Rhine Rift Valley. Several major volcanic areas rise in isolated clusters on both land and sea, including the Tibesti Mountains of Saharan Africa, the Hawaiian Islands, and many submerged seamounts (volcanoes not yet built above sea level) in the Pacific Ocean.

The vast majority of volcanoes thus lie near the oceans, and, until recently, this aspect of their distribution was rather puzzling. Equally, most volcanoes are closely associated with earthquakes, which are also concentrated in narrow bands. It had long been known that earthquakes and volcanoes were somehow related, but nobody really knew the reason until closer analysis brought some intriguing correlations to light.

The first correlation was that many earthquakes occur beneath, or near volcanoes. These are shallow, mostly less than 5 km in depth and happen with increasing frequency before, and sometimes during volcanic eruptions. Such earthquakes are weak and usually less than magnitude 4 and are often accompanied by swarms of minor tremors. (The severity of earthquakes is measured on the logarithmic Richter scale, based on the strength of shockwaves propagated through the Earth's crust.

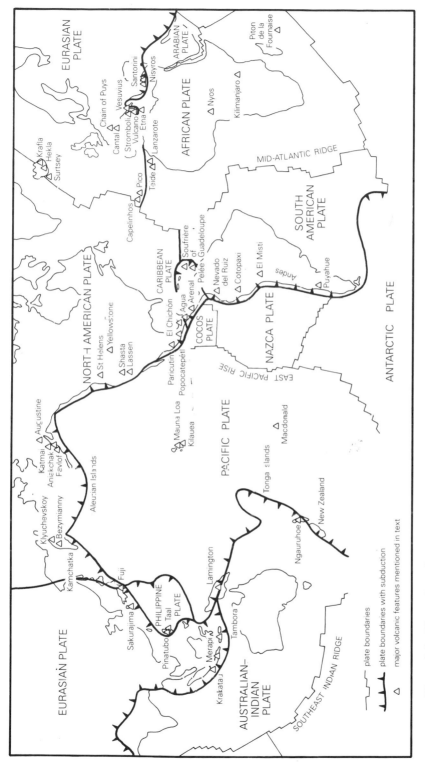

Figure 3.1 The major volcanoes mentioned in the text, with the main lithospheric plates, mid-ocean ridges and subduction zones.

Thus, a magnitude 5 earthquake is ten times greater than one of magnitude 4, and so on up to magnitude 10 or more.) Such earthquakes are closely tied to volcanoes, both in time and space, because they are generated by the rise of molten material towards the volcano itself.

The second correlation was found on the mid-ocean ridges. Earthquakes accompany the quasi-continual volcanic action along the mid-ocean ridges. Many of these earthquakes are weak, but the strongest tend to range from magnitude 5 to magnitude 6 on the Richter scale. They are also generated at a greater depth, commonly some 25 km below the crests of the mid-ocean ridges.

The third correlation between earthquakes and volcanoes occurs around the "Pacific Ring of Fire", in the Lesser Antilles of the West Indies and also, to a degree, in the Mediterranean Sea. The earthquakes occur closely parallel to the lines of volcanoes. These earthquakes are often vigorous, exceeding magnitude 7 or magnitude 8 on the Richter scale, and are extremely destructive if they happen in built-up areas, as was testified by those at Anchorage, Alaska, in 1964, in the Ionian Islands, Greece, in 1953 and at Messina, Sicily, in 1908. These are deep-focus earthquakes, generated at greater depths than any others, with their main centres often between 100 km and sometimes as much as 500 km below the Earth's surface. They occur, moreover, along sloping planes known as Benioff zones after one of the scientists who discovered them.

The volcanoes in the belts parallel to these zones of vigorous earthquakes are also marked by violent eruptions. On any individual volcano, the explosive episodes are often separated by decades or centuries of tranquillity. However, the volcanoes in these zones are so numerous that scarcely a month goes by without one of them being active. But cases are extremely rare of an eruption resulting directly from a deep-focus earthquake. (The earthquake that destabilized the north flank of Mount St Helens at the start of the eruption on 18 May 1980 was very shallow and relatively weak at magnitude 5.) The deep-focus earthquakes do not cause eruptions, nor vice versa, but both represent different aspects of the same fundamental relationship.

Curiously enough, it was increased investigation of the oceans in the 1960s that revealed the most likely connection between earthquakes and volcanoes – and even explained the frequent relationship between volcanoes and the sea. The facts known about volcanoes, earthquakes and the ocean floors were then pieced together in another fashion. It was as if the methodological kaleidoscope of the Earth sciences had been given a great jolt. A new pattern emerged which then appeared so obvious in its broad outlines that it now seems amazing that it had never been comprehended before. The distribution of earthquakes and volcanoes, instead of being the problem, actually provided the solution – and the science of plate tectonics was born.

It was discovered that the unstable volcanic and earthquake belts mark the edges of plates that together make up the lithosphere, the Earth's outermost shell, composed of the crust and the upper rigid part of the mantle. The oceanic and continental crusts form the surface layers of the lithosphere, but the plate boundaries do not correspond to the limits between the continents and the oceans. The Eurasian Plate, for example, extends westwards as far as the Mid-Atlantic Ridge (including the eastern third of Iceland), and stretches eastwards as far as Japan (but does not include most of India or Arabia). In contrast, the Pacific Ocean floor is composed of one large plate – the Pacific Plate – surrounded by several smaller plates. The surface of some plates is wholly composed of oceanic lithosphere, but others have both oceanic and less dense continental lithosphere. The plates are generally rigid within themselves, but they are in constant movement across the surface of the globe. The mid-ocean ridges mark the zones where the plate edges are pulled apart and diverge as a result of movements in the mobile layers of the Earth below. The Atlantic Ocean has one major mid-ocean ridge, running north to south approximately in mid-ocean. The Pacific Ocean, on the other hand, contains several mid-ocean ridges – notably near its eastern boundaries – representing the edges of its plates. As the plates diverge, fractures or rifts develop on the ridge crests and molten magma from the Earth's interior then ascends and leaks out from them as eruptions. The rise of this magma causes the eruptions, which are described more fully in the next chapter. When the magma solidifies as lava, it is added to the plate edge. Further divergence caused by continued movement beneath the plates generates other fractures up which more magma may then rise. Plate divergence, the ascent of magma, and its accretion to the growing edge of the plate are thus going on all the time. The plate is rather like a conveyor belt moving away from the mid-ocean ridge where its diverging edge is constantly being augmented by new magma. This is "sea-floor spreading". The newest parts of the ocean floors occur on the ridge crests, older bands run parallel to the crest on either side of it, and the oldest parts occur, as might be expected, on the outer edges of the oceans towards which they have spread farthest from the ridges where they originated. The very oldest parts of the present ocean floors are no more than about 180 million years old. Together these relatively youthful ocean floors constitute two-thirds or so of the rock surface of the globe.

Such sea-floor spreading could only be accommodated on the globe in two ways. Either the Earth must itself be expanding (which is unlikely on all sorts of other grounds), or the ocean floors must somehow be consumed at more or less the same rate as they are being formed along the mid-ocean ridges. This second view was soon verified. The true significance of the Benioff zone of severe deep earthquakes became clear. The outer, leading edges of the plates converge and one plate descends beneath another as a broad, cold, solid slab into the hot, plastic layer, known as the asthenosphere ("the zone without strength") situated below the lithosphere. This process is called subduction, and it is the main way in which the global plates are consumed. The overall speeds of subduction range from 2 cm a year to about 10 cm a year from plate to plate. The subduction can sometimes be accomplished by smooth downward sliding, but the slab usually descends intermittently both in time and space. It is the jerky downward motion, or a sudden temporary rebound, that produces the violent deep-focus earthquakes. Once subduction has been initiated by differences of temperature, pressure or density, it continues as a passive response to the sinking of the dense cold plate into the hotter zone below the lithosphere, where it is eventually assimilated at depths of about 700 km. Thus, one oceanic plate may be subducted beneath another, or an oceanic plate may be subducted beneath a plate also carrying continental crust. A continent cannot itself be subducted because it is composed of less dense material than the ocean floors and is therefore buoyed up (and hence the continental rocks can be very much older than those of the ocean floors). (Although it is not directly related to volcanicity, plates sometimes also neither converge nor diverge, but slide past each other along major fractures or transform faults.)

The zone where subduction begins is marked by long trenches, which are commonly between 5000 m and 11 000 m deep. Such trenches, the deepest part of the oceans, are deeper even than the abyssal plains that form most of the ocean floors. The angle at which the plate is subducted varies according to many factors including the relative densities, temperatures and mineral contents of the slab, compared with those of the materials into which it is sinking. The Marianas subduction zone, for example, curves down until it is almost vertical, whereas the subduction beneath the Peruvian Andes takes place at an angle of some 15°.

Subduction initiates a series of complex changes, described in the next chapter, which result in partial melting of the asthenosphere and the formation of magma. In subduction zones, the magma rises intermittently to the surface, with sojourns of varying lengths in reservoirs en route, where it undergoes changes that usually cause it to erupt violently above the zone of melting. The volcanoes often form on the overriding plate about 150–200 km from the ocean trench marking the inception of subduction – the distance depending primarily on the angle at which the plate is subducted. They form lines of islands in the oceans or chains on land, which are frequently more than 1000 km long and about 50 km wide, with volcanoes spaced at intervals ranging from 10 km to 100 km apart.

When an oceanic plate is subducted beneath another oceanic plate, the resulting volcanoes rise from the ocean floor to form a gently curving island arc. These may be relatively simple volcanic island series such as the Aleutian arc, or they may have a greater age and complexity derived partly from the addition of continental masses into their make-up such as Japan, Indonesia or New Zealand. When an oceanic plate is subducted beneath a continent-carrying plate, the magma rises up fractures through the continental crust onto the surface and creates land-based, usually straighter, volcanic chains represented by the Cascade Range of the United States and Canada, or the volcanoes in Central America and the Andes. (When two plates carrying continents converge, as occurred beneath the Indian and Eurasian Plates, the collision develops abnormally thick continental crust, found, for example, in the Himalayas and Tibet, but little volcanic activity directly results.)

The magma generated by subduction does not reach the Earth's surface continuously, nor at the same rate, otherwise the volcanoes in a given subduction zone would all erupt in a prolonged explosive chorus. Two outbursts along the same subduction zone at the same time are, in fact, rare. The eruptions in the West Indies of the Soufrière on St Vincent on 7 May 1902, and of Montagne Pelée in Martinique on the following day, were quite exceptional – at least as far as historic records and absolute dating methods have revealed. Because the magma that rises from subduction zones often remains in reservoirs for a time, eruptions from subduction zone volcanoes are irregular and separated by long dormant periods. Moreover, in any one chain or arc, some volcanoes become extinct long before others. In the Cascade Range, for example, Three Fingered Jack has long been extinct, but Mount St Helens is clearly not.

The mid-ocean ridges where the oceanic plates are created, and the subduction zones where they are consumed, account for most of the world's volcanoes. But not all. Two relatively small sets of volcanoes associated with continental rifts, and hotspots on land and sea, together account for the remaining 5 per cent of the world's active volcanoes.

Rift volcanoes occur along, or near, great parallel faults which transect some of the continents, where they have been broken apart when the lithosphere stretched in response to plate movements. Some rifts may represent either the beginnings of future oceans, such as the East African Rift system, or the remains of aborted ocean-forming processes, such as the Rhine Rift Valley.

Hotspot volcanoes are intriguing exceptions in the volcanic world. Whereas most volcanoes are formed by plate movement, hotspots occur where plumes of hot plastic materials rise up from the asthenosphere, or perhaps farther down, and generate magma. It eventually bursts through the lithosphere to form volcanoes in both the oceans and on the continents such as the Hawaiian Islands, the Galápagos Islands, Réunion, and the Hoggar and Tibesti Mountains of the Sahara. Because the plumes remain fixed in position (or move only very slowly) and the lithospheric plates travel over them, lines of volcanoes are formed that, in fact, can also indicate the direction of plate movement. Thus, the Hawaiian Islands were formed by the most recent eruptions at the southeastern end of a line of older, often submerged volcanoes, which increase systematically in age northwestwards. This demonstrates that the Pacific Plate, through which the lavas have burst, has been gradually drifting northwestwards to subduction in the Aleutian and Kurile Trenches.

The origin and distribution of volcanoes is now therefore much better understood than a few decades ago. All these features occurring along linear bands – volcanoes, earthquakes, trenches, rifts and even, indirectly, rift valleys and hotspots – can be fitted together in a pattern which provides not only the broad explanation, but also the mechanism generating them. Volcanic eruptions are thus intimately related to the main single process affecting the Earth's lithosphere: the growth and the destruction of its component plates.

4

Generation and ascent of magma

The explanation of the distribution of volcanoes shows that what happens to materials deep underground usually determines their site, contents, behaviour and appearance. These subterranean processes generate magma, which rises to the surface, erupts, and creates volcanoes. Magma is hot, mobile, molten rock material melted from parent layers below the Earth's crust, and which also contains crystals in suspension and dissolved gases, or volatiles. Magma erupts at the surface as lava, in flows or fragments, and as gases. Depending on their composition, the temperatures of typical magmas near the Earth's surface range from about 1200°C to about 700°C.

The interior of the Earth

The Earth is composed of a dense core, surrounded by a series of concentric shells, each with different chemical and physical properties (Fig. 4.1). The crust, forming the Earth's surface both on the continents and on the ocean floors, is the only cool and solid shell. Most of the materials beneath it are very hot and held under great pressures. The nature of the Earth's interior has been inferred partly from clues prised from volcanic rocks themselves, and partly from small fragments, known as xenoliths, which are sometimes erupted onto the surface from the depths. However, the most valuable information has been derived from the science of geophysics, most notably through studies of the ways that seismic waves are propagated during earthquakes.

The Earth's radius is 6400 km and the core, which makes up one-third of the global mass, has a temperature of about 4300°C. The inner core is believed to be hot and solid because of the enormous pressures that confine it, and seems to be composed of iron. The outer core is thought to be in a fluid state and also made up largely of iron, with additional nickel and a few other lighter elements. The core is surrounded by the mantle. The solid lower mantle, or mesosphere, is enveloped by a transition zone, which, in turn, is surrounded by the upper mantle. The transition zone and the plastic part of the upper mantle are together known as the "asthenosphere", the zone without strength. All these mantle zones are largely made up of silicate minerals. The outer part of the upper mantle is solid.

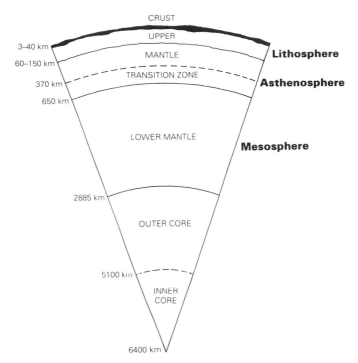

Figure 4.1   The internal structure of the Earth.

At the high temperatures and pressures prevailing within the mantle, stresses and strains, various forms of creep, as well as convection develop. The consequent movements and deformations within the mantle are usually slow, but are probably continuous, especially in the asthenosphere, its weakest and most mobile layer, which plays an important rôle in generating magma. The asthenosphere reacts to stresses in a fluid way, although it is only partly molten.

The outermost shell of the Earth is the lithosphere. Its base comprises the solid outer parts of the upper mantle that are capable of plastic flow, and its surface layer forms the cool, solid and brittle crust of the Earth. The lithosphere is between 100km and 150km thick under the continents, but is only between 60km and 70km thick below the oceans. The crust itself reaches an average of about 40km in thickness on the continents (and as much as 70km under mountain ranges), but it can be as thin as 6km on average beneath the ocean floors (and even less than 3km thick under the mid-ocean ridges). The continental crust is pale and broadly granitic, or silicic, in character, and has a low density, whereas the oceanic crust is dark and basaltic in character, and has a higher density. The whole lithosphere – including the crust – is broken into a dozen or more, slowly moving, rigid plates, as was described in the last chapter.

The formation of magma

Magma is created by partial melting of the asthenosphere, which, although it varies to some extent from place to place, is largely composed of the olivine-rich rock peridotite. Only a fraction melts because only some of its mineral constituents reach their melting points, whereas others do not, and therefore remain in a solid state. The melt, often called primary basaltic magma, is thus not the same as the asthenosphere from which it is derived. Magmas of different compositions can later develop from primary magmas. However, all magmas have three components: molten material composed largely of silicates; unmelted crystals; and contained volatiles, gas or steam, which separate out only when the magma closely approaches the surface.

Partial melting can result from a number of different processes, such as increases in temperature, or reductions in pressure as displacements occur. Temperatures may increase if a mass sinks lower into the mantle, or if it is heated when hot materials rise from below. Pressures would be reduced, and melting could then take place, if a mass were to rise closer to the surface. Melting may also be induced if water is added, which most often happens when hydrous mineral assemblages in a subducted plate are heated. Water is released and then moves into the asthenosphere and lowers the melting temperatures of some of its constituents.

The magmas generated fall into two broad chemical types: basic and silicic. There are, however, other ways of describing and classifying magmas and there are many more types of magmas than those mentioned here. Moreover, as research has progressed, ideas about magma generation have also changed: they are the subject of detailed study in the science of petrology.

Basic magmas

The chemical composition of basic magmas differs least from that of the mantle from which they originate, and it therefore seems likely that they transit relatively quickly to the surface. Basic magmas are typically dark, heavy and dense. They predominate in eruptions in both mid-ocean and hotspot environments, although they are also common in subduction zones. The most common kind of basic volcanic rock, basalt, contains a relatively modest amount of silica (between 45% and 54%) but it is rich in ferro-magnesian constituents. The typical minerals of a basalt are calcium-rich feldspar and ferro-magnesian minerals such as olivine, pyroxene and magnetite. Olivine and pyroxene, in particular, are also commonly found in xenoliths. Basalts vary partly because they are formed in different positions within the asthenosphere. Tholeiitic basalt is the most silicic type, with about 50 per cent silica and only 3 per cent alkalis, for example; olivine basalt has a rather lower silica content; and alkali basalt is richer in both potassium and sodium (4%), but has only about 45 per cent silica. Most basalts, however, share a common characteristic: they erupt hot and fluid. The notably fluid basalts often expelled by Etna and Kilauea, for instance, erupt at temperatures ranging between 1050°C and 1200°C.

Silicic magmas

For the sake of simplicity here, "silicic" magmas are grouped with "intermediate" magmas. Silicic magmas are paler in colour, lighter in weight, less dense and, under the same temperature and pressure conditions, are usually more viscous and explosive than basic magmas. Silicic magmas erupt in much smaller volumes than their basic counterparts, even in the subduction zones where they are most commonly generated. They can be broadly subdivided, with blurred boundaries, according to their silica contents, which are higher than those of basalts. Andesites contain about 54–62 per cent silica, dacite contains between 63 per cent and 69 per cent silica, and rhyolite has between 70 per cent and 78 per cent. The notably viscous obsidian is the glassy and dark-coloured equivalent of rhyolite. Silicic magmas are cooler than basaltic magmas when they emerge. Intermediate andesites range in temperature from 900°C to 1000°C – the temperature of the lavas erupted by Parícutin in 1943. Dacites erupted from Mount St Helens at temperatures of 850–930°C, and rhyolites can emerge as cool as 700°C. Their relatively high silica contents have been attributed to evolution from basalts in a reservoir, to melting of siliceous rocks in the crust, or to remelting and differentiation of previously generated basalts. These processes also lead to the formation of trachytes and phonolites, which have high silica contents, between those of andesites and dacites, but are also richer in sodium and potassium than silicic magma.

Magma viscosity, volatile content and bubbles

Viscosity is a complicated subject, where simplification almost inevitably leads to half-truths. It deserves consideration, however, because viscous magmas generally erupt violently. Viscosity results from the interaction of such factors as temperature, pressure, volatile content, chemical nature, and the contents of crystals and volatile bubbles. The molecules in the magma are polymerized, or banded together in clusters. Silicic magmas are more polymerized, and thus more viscous than their basic counterparts. It is perhaps the silicon–oxygen bonds, and to a lesser extent the aluminium–oxygen bonds, which are the strongest cation–anion bonds in a magma and the main influences on its viscosity. These bonds play a large part in giving a clear structure to a magma, even when it is still too hot to crystallize. When a magma cools as it approaches the surface, a bonded, crystalline structure develops, which increases its viscosity. But, on eruption, basaltic magmas have developed only a small proportion of crystals, whereas they can make up half the volume of silicic magmas.

The main volatiles in magma are water, carbon dioxide and sulphur dioxide, which rise to the upper parts of the magma when they separate out as it nears the surface because of their lower density. Although water represents only a very small proportion of a magma, it nevertheless plays an important but far from simple rôle. Water is more soluble in cooler magmas, and thus cooler silicic magmas often contain more water (up to 5%) than basaltic magmas (which contain about 1%). On the other hand, the presence of dissolved water lowers the viscosity of magmas, because its ions – especially the vigorous hydrogen ions – break down the crystal polymer bonding networks within the magma. But, near the surface, as the water vapour separates out and forms bubbles, the overall viscosity of the magma may actually increase. This process is sometimes likened to the stiffening that occurs when the white of a raw egg is whisked and bubbles form.

Thereupon, another physical factor comes into play. As the magma rises closely towards the surface, the pressures upon it are reduced and more and more bubbles can then separate out from its still-fluid parts (Fig. 4.2). If the volatile content is so high that many bubbles develop, and thus make the magma very vesiculated (or frothy), then the stiff films of magma forming the walls of the bubbles are thinned and weakened so that the bubbles can expand – often greatly increasing the volume of the magma as a whole. When the magma erupts, sudden surface decompression enables the bubbles to burst in violent explosions that shatter the magma into small fragments that then cool quickly. Near the surface, basic and silicic magmas behave differently, often because of their different viscosities and volatile contents. Thus, because the walls of bubbles in fluid basaltic magmas are relatively weak, they can burst more easily, less violently, and over a longer period. Moreover, many basalts are so fluid that bubbles can move through them and reach their surface without any explosions at all. But the properties of silicic magmas often conspire to generate sudden, violently explosive eruptions, especially where the volatile content predominates. Such eruptions best display the vigorous expansion of volatiles that transforms the thermal energy of the magma into the kinetic energy that can sustain a vast eruption column. However, when most of the volatiles have been expelled (often when an eruption is waning), viscosity may then predominate and domes may be slowly extruded. Thus, once Mount St Helens had disgorged its gas-charged columns of fragmented magma in 1980, its eruptions during the ensuing six years were in a much more minor key and were devoted to the creation of a small dome.

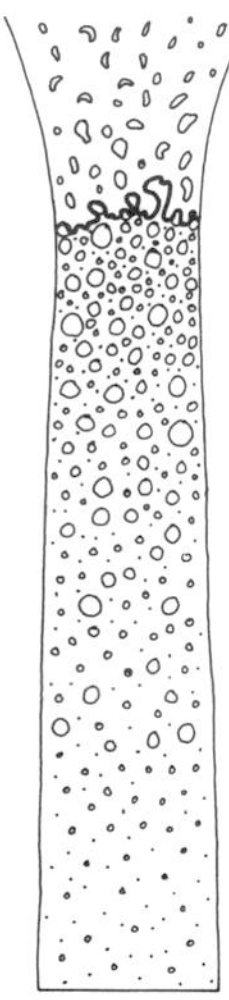

Figure 4.2
Bubble formation in a magma.

Eruptive environments

The mantle has a density of about 3.3; the density of the hot melts forming magmas is about 2.9. Thus, magmas are relatively buoyant and they tend to rise towards the Earth's surface; they erupt when they overcome the confining pressure of the lithosphere. The distribution of volcanoes shows that magma rises towards the Earth's surface in four main environments: on mid-ocean ridges, in subduction zones, on continental rifts, and at hotspots. Most volcanoes are either on mid-ocean ridges or in subduction zones (Fig. 4.3). There is a correlation between the nature of the rising magma and the position and behaviour of the volcanoes in each environment.

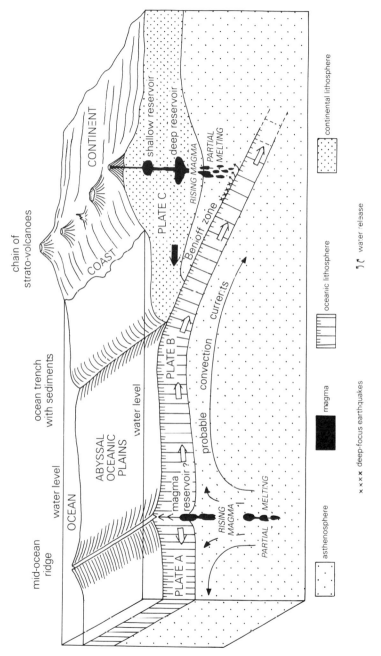

Figure 4.3　The growth of volcanos at mid-ocean ridges and subduction zones (not to scale).

Ascent of magma on mid-ocean ridges

From the point of view of volcanic eruptions, the mid-ocean ridges constitute a comparatively simple environment. Here the asthenosphere, largely composed of peridotite, at temperatures usually between 1200°C and 1300°C, makes its closest approach to the surface, and the cool lithosphere can be as little as 3 km thick. As the plates diverge, the hot peridotite rises towards the crest of the mid-ocean ridges, where the pressures upon it are reduced so that it begins to melt. However, at most, only 30 per cent of the peridotite can ever melt, because nearer the surface the temperature is too low, and farther down the pressure is too high. At depth, the high temperatures would facilitate melting, but it is prevented by the great pressures exerted there, whereas, nearer the surface, lower pressures would allow more melting, but it is inhibited by the prevailing lower temperatures. Thus, only some of the constituents of the peridotite melt out and the primary basaltic magma so formed is not the same as its parent.

The primary basaltic magma is less dense than the peridotite and can rise towards the surface because it is more buoyant. Moreover, as the lithospheric plates diverge, vertical fractures, extending from the asthenosphere to the surface, are generated running parallel to the mid-ocean ridges. Magma penetrates these cracks in vertical sheets (or dykes), but it will only erupt onto the surface if the magma pressure is greater than the minimum principal stress in the crust, because the overall lithospheric stress field controls the overall rate of magma ascent. Thus, although magma here rises relatively quickly *towards* the surface, only a proportion of the rifts actually disgorge (chiefly tholeiitic) basalts *onto* it. The rest – probably over half the volume of the rising magma – solidifies in the fractures and becomes welded to the edges of the diverging plates. These features have been well studied where the Mid-Atlantic Ridge emerges in Iceland.

Although the magma often rises rapidly (in the geological sense) to the surface, it sometimes halts a while in shallow reservoirs where important changes in its composition and mineral content may take place. For example, more silicic rocks may develop if hydrated basalts are remelted and evolve in the roots of the dykes. Hence rhyolites have been emitted from Askja and Hekla in Iceland.

When the magma erupts onto the land surface, cinder cones and extensive basaltic flows are formed along fissures. In most cases, however, eruptions take place frequently all along the submerged crests of the mid-ocean ridges, where they form thick accumulations of pillow lavas which are added to the margins of the diverging oceanic plates and eventually conveyed away from the ridges. Thus, basaltic flows, and especially lava-filled dykes and pillow lavas comprise the main constituents of the ocean floors beneath their thin cover of sediments.

Ascent of magma at hotspots

Hotspot volcanoes are generated by hot rising plumes of plastic rock material formed deep within the mantle, perhaps even near the core, which are probably caused by thermal or compositional anomalies that are thought to be allied to convection. The rising mantle plumes cause some melting of the asthenosphere starting at depths exceeding 50–60 km and they form alkali basalts. As the plume continues to rise, further melting can also generate tholeiitic basalts similar to those erupted on mid-ocean ridges. Thus, although mid-ocean ridges usually give rise to tholeiitic basalts, hotspots usually also produce alkali basalts. At Kilauea on Hawaii Island the magma is continually fed into a shallow reservoir, 3–6 km deep, and escapes from time to time along a series of vertical dykes into which it is inserted like a knife blade. The progress of the magma is manifested (and can be monitored) by a multitude of small earthquakes, and, if the magmatic pressure is greater than that exerted by the walls of the dykes, then an eruption will take place. The overall result is a vast shield, predominantly composed of lava flows, which in Hawaii Island itself rises 9000 m from the Pacific Ocean floor.

Some continental areas have also shown activity, probably related to mantle plumes, which has given rise to many copious basaltic outpourings, such as the Snake River Plains of Idaho. These basalts can also be either tholeiitic or alkaline in character, but they are sometimes accompanied, as in Mont-Dore in Auvergne, by more evolved alkaline rocks, such as trachytes, or rhyolites, which were probably formed after the magma halted in a crustal reservoir.

Ascent of magma in continental rifts

Volcanic activity in continental rifts is associated with stretching and thinning, updoming and the development of deep fractures in the lithosphere up which the magma may then rise. Such magmas are usually alkaline and range from alkali basalts, where ascent is quite rapid, to more evolved alkaline magmas such as trachytes and phonolites. In the Grande Limagne Rift in Auvergne, for instance, extensive alkali basalt emissions occurred in the west, often from vents along deep fissures, whereas more restricted eruptions of more viscous phonolites were expelled in the Comté d'Auvergne in the east.

Ascent of magma at subduction zones

Subduction zones are the most complex areas where magmas are generated (Fig. 4.3). These are the zones of explosive silicic volcanic activity, which is the antithesis of the quiet effusions common amongst basaltic magmas. The exact origin of magmas in subduction zones has always been the subject of debate. It was thought at first, for instance, that the magmas were mainly derived from melting of the upper surfaces of the subducted plate, but closer petrological analyses indicated that they were more probably derived from partial melting of peridotites in the wedge-shaped mass of the asthenosphere situated above the subducted plate and below the overriding plate. Subsequent isotopic studies, however, suggested that continental sediments must also play a rôle. They would be deposited in the ocean trench, then subducted with the descending plate, before being eventually taken up in the asthenospheric melting, or they could be derived from some melting of continental crust. As the plate is subducted, different types of magma are generated which are largely determined by the pressure and temperature conditions prevailing at different depths. More magma is also generated when the subducted plate is shallow than when it has sunk deeper. Thus, in broad terms, the resulting eruptions can change with time as the plate is subducted farther. These temporal changes are sometimes reflected in the landscape, when different magmas can erupt in adjacent bands.

When a subducted plate sinks to a depth of more than 80 km, melting commences in the asthenospheric wedge above it, as now occurs beneath young island arcs such as the Tonga–Kermadec, the Aleutian Islands, and parts of the Lesser Antilles arcs, where subduction has not lasted long enough to reach great depths. Partial melting starts with the successive transformation of certain minerals within the descending plate, and especially when water is released, as subduction progresses downwards and pressures increase. Thus, amphibolite in the descending plate changes to the basaltic-type rock, quartz eclogite, under the high pressures pertaining at depths of about 80–100 km. This releases water, which then rises into the asthenospheric wedge and lowers the temperature at which partial melting of the peridotites there can take place. The melt is less dense than its surroundings and can thus rise – and eventually erupt – as tholeiitic basalts, basaltic andesites, or andesites.

As subduction continues into zones of even greater pressure at depths between 100 km and 150 km, the transformation of other minerals such as serpentinite also gives off more water, which enables further melting to occur. At these depths, too, the quartz eclogite partially melts as a result of the high water vapour pressure, and, in turn, causes even more asthenospheric peridotite to melt in the wedge above. As the buoyant melt rises into the lithosphere, complex processes may then generate silicic magmas, which are predominantly andesites, but often also dacites and rhyolites, as well as some basalts. Such silicic magmas are a feature of older, mature volcanic arcs, such as those of Japan, Indonesia or New Zealand, where the overriding plate often contains some continental crust, about 20–35 km thick.

Where an arc has developed for millions of years, such as, for example, in Indonesia, its older, deeply subducted parts in the west display violent explosions; the younger, more recently subducted zones in the east are still erupting tholeiitic basalts. The Tonga–Kermadec–New Zealand arc provides even more striking contrasts. The Tonga–Kermadec sector is young, narrow, mainly tholeiitic and rests on oceanic crust. On the other hand, the New Zealand sector is older, wider, and rests on some continental crust. Although andesites and basalts occur in New Zealand, they are dominated by about 16000 km^3 of silicic rhyolite, which was probably formed in association with partial melting of continental crust. Whatever the reason, silicic magmas are not only more diverse but also more voluminous where continental crust occurs on the overriding plate.

Where subduction lasts even longer, and the descending plate sinks below a depth of 200 km, melting in the asthenospheric wedge develops alkali basalts, and eventually trachytes and rhyolites. However, the exact causes of the development of these magmas are still to be resolved. In Japan, where the volcanic arc is old established and unusually wide, the three kinds of magmas erupt in almost parallel belts. Tholeiitic magmas, including those of Fuji, rise nearest the ocean trench, where the subducted plate is shallow; silicic magmas, including those of Unzen, rise in the middle zone; alkali basalts rise farthest from the ocean trench where the subducted plate is deepest.

The magmas generated in subduction zones acquire even greater variety as they ascend (Fig. 4.4). Because they are less dense than their surroundings, they rise in

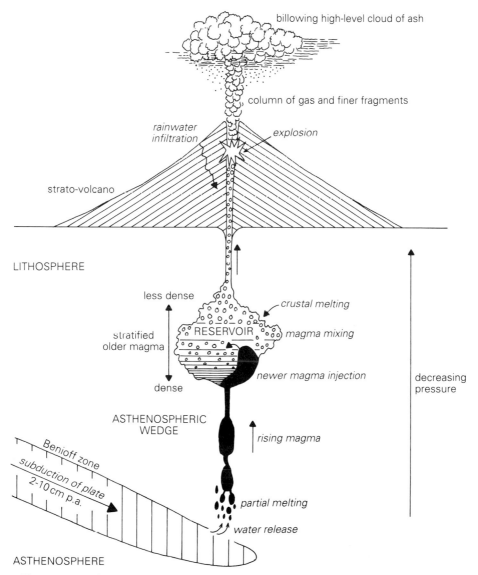

Figure 4.4 The main factors that may be involved in an eruption from a volcano in a subduction zone (not to scale).

broad, globular, buoyant masses called diapirs ("that which cuts through"), which may be able to thrust aside crustal rock layers, and can also melt their way upwards by melting their roofs. Many diapirs, however, fail to reach the surface before cooling and slowly solidifying below ground, and are then only exposed after the overlying crust has been much eroded. But others frequently cause volcanic eruptions above subduction zones.

During their slow ascent through the lithosphere, magmas often penetrate into chambers or reservoirs. A reservoir is not a hole or infernal cavern, but probably a highly fractured zone within, or just below, the overriding plate. Major reservoirs in subduction zones are commonly from 25 km to 50 km deep and some 20 km³ in volume, and they may retain magmas for many centuries. They are often relayed by smaller, much shallower reservoirs, often ranging between 3 km and 10 km in depth, and 2 km³ and 5 km³ in volume, where magma is only usually held for much shorter periods.

The major reservoirs mark the sites of significant changes in the nature of the magma. Generally speaking, the longer the magma halts in a reservoir, the more it differs from the original melt. The magma evolves and differentiates by means of interrelated factors such as cooling, fractional crystallization, and gravitational stratification, and it may also be contaminated by melting from the walls of the reservoir. These changes develop magma which is usually more viscous, more gaseous, and therefore commonly much more explosive when it erupts.

As magma enters a reservoir its temperature may be anywhere between about 1000°C and 1400°C. It soon begins to cool and some of its component minerals crystallize out, starting, of course, with those with the highest melting (or solidification) points. This partial, or fractional, crystallization means that the still-molten magma gradually changes in composition, and a partially crystallized mush is formed. The crystals separated early on sink towards the bottom of the reservoir, essentially because the denser ferro-magnesian minerals have the highest melting points. The remaining molten magma, having lost many of its ferro-magnesian constituents, thus becomes more silicic. Furthermore, since the minerals that separate out early on do not include volatiles in their atomic lattices, the molten magma that remains becomes progressively richer in volatile constituents, especially in water. All these factors tend to make the reservoir become more and more stratified and more evolved.

Thus, magmas that were originally basaltic or intermediate change into more silicic, more viscous and more volatile magmas as their evolution proceeds. In this way, for example, basalts may eventually evolve fractions which are andesitic, or dacitic, or even rhyolitic in character. At the same time, however, the magma may melt part of its confining walls and become contaminated in various ways, depending on the nature of the melted walls, which could, for instance, range from basalt to granite in composition.

Still further changes within the reservoir may be induced when hotter magma wells up and mixes with the more evolved magma already installed to form hybrids with a composition intermediate between the two. For example, less than 1100 years ago, the reservoir beneath Glass Mountain on the Medicine Lake volcano in northern California was full of molten rhyolite. It was then invaded by basalt from below, which mixed with some of the rhyolite to form a layer of dacite that settled above the rhyolite. The arrival of the basalt, however, triggered off an eruption, which first expelled the dacites, then a hybrid mixture of rhyolites and dacites, and finally the rhyolites from the lower parts of the reservoir. Such upsurges of new magma often unleash eruptions. An incursion of hot basalts into a reservoir containing cooling dacites, for example, probably caused the eruption of Pinatubo in June 1991.

The evolved and dilated magma that resumes its rise towards the surface above a subduction zone is thus in a very different state from when it started on its journey. Once the buoyant magma begins to approach the surface, an eruption becomes almost inevitable, because considerable magmatic pressure develops, at the same time as confining crustal pressures upon it are reduced. Within about 2km of the surface, decreasing crustal pressures enable the volatiles contained in the magma to separate out. The bubbles so formed not only increase quickly in number and size, but they also rise through the mass and help to propel the magma more rapidly upwards through the last stages of the journey (Fig. 4.2). Events thus happen at an increasing pace. When the magma may already lie within the structure of the stratovolcano, the bubbles can become so numerous that the whole upper mass of magma becomes a frothy, foam-like liquid. The volatiles suddenly overcome both the viscosity of the magma and the now much-reduced crustal pressures. The gases expand extremely rapidly. The cork is blown from the champagne bottle. The volcano erupts. Violently. The magma is pulverized into fine, suddenly chilled lava fragments of ash, pumice, and volcanic dust that are thrown by the enormous gas discharge high into the air in a billowing column. As more magma rises to the surface, the bubbles repeatedly expand with great violence over a period that can last as long as 48 hours in a major eruption. Huge amounts of energy are released which are much more powerful than any nuclear bomb yet exploded. The magma that erupts first – and precipitates the violent initial blasts – is that with a high volatile content from the summit layers of the reservoir. This sudden explosive evacuation sometimes causes the roof to collapse into the chamber, and gives rise to a further extremely violent caldera-forming eruption, as occurred, for example, at Santorini and Krakatau. Eventually the magma supply is exhausted and none remains to sustain the eruption. The volcano then lapses into a long dormant period during which the magma may be replenished and the whole process can begin all over again.

5

Eruption styles

Eruptions often make volcanoes notorious, but few cause cataclysms that destroy vast areas and kill thousands of people. In fact, many eruptions take place unseen on the ocean floors and others have merely given rise to swarms of (often nameless) little cones. Only a few outbursts in every century cause big explosions and decapitate mountains, but, on the other hand, emissions repeated for a million years or more can accumulate great volcanic volumes on hotspots and mid-ocean ridges. The transfer of material from the depths to the Earth's surface thus takes on many different guises. The consequences are seen in cones, domes, lava-flows, blankets of ash, nuées ardentes, debris avalanches, lahars and calderas, whereas altogether weaker emissions give off gas, geysers and boiling mud from sulphurous pits. Not every eruption creates a Vesuvius.

The nature of every volcano is determined by its predominant eruptive styles, which are themselves determined by the kind of magma rising to the surface. The possible combinations of eruptive styles and resulting landscape forms may seem almost infinite, but it is, nevertheless, still possible to distinguish major kinds of eruption, each of which have many gradations in between them. At the beginning of this century, four main groups were singled out, represented by the eruptions observed on Hawaii, Stromboli, Vulcano and Montagne Pelée (Fig. 5.1). But many eruptive styles, and many volcanoes, did not fit readily, if at all, into this four-fold division, and it was observed that even the archetypal volcanoes had produced different types of eruption. Moreover, new techniques and observations revealed types of activity that had previously been neglected or unknown. To cite but two examples, the study of the mid-ocean ridges has shown the importance of submarine emissions and the eruption of Mount St Helens emphasized the rôle of debris avalanches. Thus, many different eruption styles, often merging without clear boundaries, are required to encompass the complexities of the real volcanic world.

Submarine eruptions

Most of the Earth's volcanic eruptions occur under the sea, chiefly emerging from vents aligned along fissures on the crests of the mid-ocean ridges. Such eruptions are dominated by the high pressure exerted by the overlying sea water, which stifles any

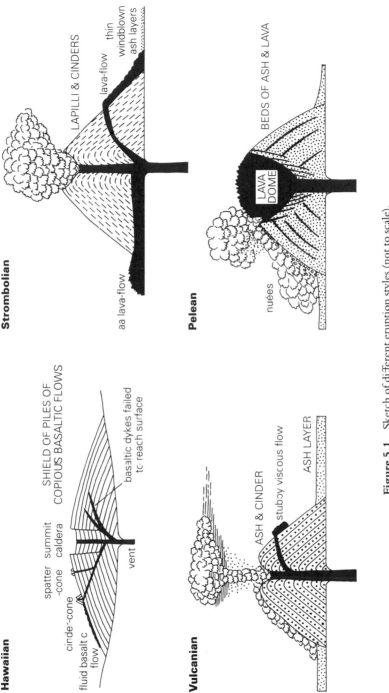

Figure 5.1 Sketch of different eruption styles (not to scale).

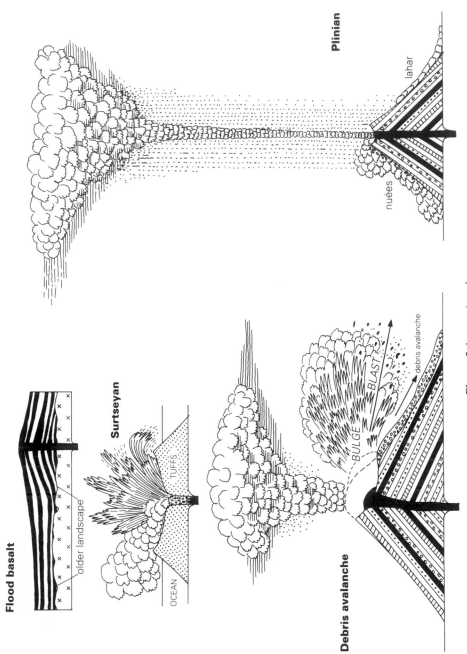

Figure 5.1 continued.

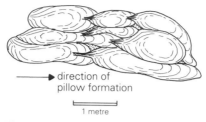

direction of
pillow formation

1 metre

Figure 5.2   Formation of pillow lavas.

potential explosions so that the basaltic lava only succeeds in oozing calmly from the vent, even at depths of a mere 100m. It cools rapidly and forms pillow lava, composed of solidified lumps the size and shape of pillows (Fig. 5.2). As the basalt squirts out, a skin soon cools around it as it comes into contact with the cold water. Continued basaltic injection stretches the skin, which swells up into a pillow form, usually less than 1m across. Further injection may burst the skin and form another pillow attached to the first. When cooling makes the skin too thick to stretch any more, the injection stops and the lava congeals inside the pillow. The basalts then find another outlet nearby and make another pillow. This calm emission of basalt all along the mid-ocean ridges is the least visible, most continuous, most voluminous and the most important volcanic activity in the world, generating new oceanic crust and helping widen the ocean floors as the plates diverge. It represents over three quarters of the Earth's magma output. But very few scientists have ever seen it happening and it had scarcely even been suspected until the 1960s outside the few areas where, as in Iceland, the mid-ocean ridges reach above the waves. These fissures also give off hot springs and gases such as hydrogen sulphide and carbon dioxide, which nourish a variety of strange life forms, and the metal sulphides associated with them, called "black smokers", discovered in 1977 off the Galápagos Islands in the eastern Pacific Ocean, may lie at the very origin of ore deposits.

Submarine eruptions also occur in isolated vents, sometimes related to hotspots, which form seamounts rising 2000m or more from the ocean floors. These, too, are formed by prolonged mild eruptions that give out great volumes of basalt, usually as pillow lavas, but they arise from concentrations of vents rather than from long fissures. If activity continues long enough, a volcanic island may emerge. Many volcanic islands, including Hawaii, Stromboli, Vulcano as well as the Azores, began their lives as seamounts. Many others are dotted about the oceans and are particularly common in the South Pacific Ocean.

Macdonald Seamount, near Pitcairn Island in the Pacific Ocean, was first discovered in 1967 and its emergence as a new volcanic island cannot be long delayed (Fig. 3.1). It now rises to within 27m of sea level, and, in October 1987, had the impertinence to erupt beneath an American oceanographic research vessel, giving off large gas bubbles and scattering hot, glassy volcanic rocks into the air.

The submarine volcano of Kick'em Jenny, 7.5km north of Grenada, is the most active vent in the Lesser Antilles volcanic arc at present (Fig. 5.3). It has erupted 10 times in the past 50 years, whereas the emerged volcanoes in the arc have only man-

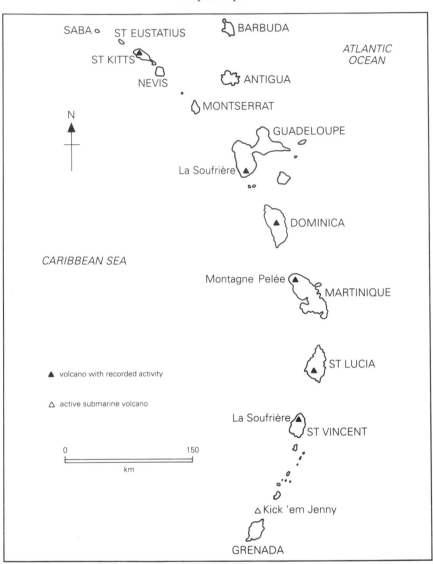

Figure 5.3   The Lesser Antilles volcanic arc.

aged seven eruptions between them since 1900. Kick'em Jenny erupts relatively small amounts of basalt derived from a depth of about 80km, rather than the more evolved andesites and dacites typically expelled by its violent insular neighbours.

Kick'em Jenny rises from a depth of 700m on the submarine ridge associated with the Lesser Antilles volcanic arc. It was first noticed in 1939, when columns of steam rose above sea level. The sea also bubbled during the eruption of 1974, but all the other eight eruptions since 1939 have failed to break to the surface of the water. In 1962 its crest lay at a depth of 232m, and by 1972 it had built up to a depth of

190 m and a crater 15 m deep and 180 m across was noted for the first time. An eruption in 1977, however, filled the crater and raised Kick'em Jenny to within 160 m of sea level. Strong earth-tremors felt on the coast of Grenada betrayed three eruptions on 29 and 30 December 1988, but nothing rose above sea level. However, at the present rate of accumulations, Kick'em Jenny seems likely to emerge in a Surtseyan eruption within the next 200 years.

Surtseyan eruptions

Surtseyan eruptions were observed at Capelinhos in the Azores from 1957 to 1958, but it was the emergence of Surtsey, off southwestern Iceland between 1963 and 1967, that gave them their name (Fig. 5.1). They are most commonly generated by basaltic magmas. Surtseyan and phreatomagmatic eruptions constitute the two main forms of hydrovolcanic activity, and they have often been grouped together. The best displays of Surtseyan activity take place mainly in lakes, shallow seas or, occasionally, above shallow aquifers. The pressure exerted by deep water that inhibits explosions during submarine eruptions is much reduced when the water is shallow. Hydrovolcanic explosions can easily develop where the water is less than 100 m deep and, when water invades the vents up which magma is rising, the eruptions are more violent than if they had taken place on land. Surtseyan eruptions are most often generated by basaltic magmas associated on land with mildly explosive activity. The sudden transfer of heat from magma at a temperature of 1200°C to water commonly at less than 20°C transforms the water into steam; its expansion releases energy that quenches and shatters the magma. Large volumes of water can continually gain access to the vent. The explosive expansion of the steam thus takes place near the surface in an open vent where water pressures and temperatures are relatively low, and where the energy generated can be dispersed in eruptions, of varying intensity, repeated at intervals of several minutes, that may last for many months. The upper parts of the conduit are filled with a fluidized slurry of fragments and water close to boiling point, that is repeatedly penetrated by pulsations of magma, each of which provokes an explosion as the water expands into steam. The shallow explosions and limited collapse of the uppermost walls of the vent develop a funnel-shaped conduit, open to the air or water at the top. Whether there is any deep foundering in the conduit is still debated. The shattered fragments of magma accumulate around the vent in mainly fine and thin beds, known as "tuffs", which are characteristically tiny, often 1 mm across, angular and glassy, because of the shattering and quenching processes in their formation. Little of the country rock is exploded, new magma usually comprising over 90 per cent of the fragments ejected.

The explosions commonly take on three main forms, which may occur together or separately. Fragments are expelled upwards in a cloud of ash and steam that can form a billowing column reaching 5 km in height. Many fragments fall back into the vent, and are shattered by more explosions until they are carried away by the wind, or finally accumulate around the orifice. The second form of expulsion occurs when the column collapses, or clouds of dry fragments and superheated steam surge from the conduit and radiate from the vent like the billowing collar developed around the bases of nuclear explosions. But the third form is the chief distinguishing mark of Surtseyan eruptions. Thick dark-pointed jets of fragments, often headed by bombs trailing black debris behind them, are repeatedly shot out from the vent. They resemble pointing fingers, the fronds of a cypress tree, or the spreading plumes of a cockerel's tail as they arch out about 1 km from the vent. These jets are usually generated by relatively cool eruptions where the trapped steam condenses around and wets the fragments, which then accumulate in consolidated layers in tuff cones.

Because water is continually available in a lake or the sea, Surtseyan eruptions continue until the magma stops rising, or until an impermeable mass of fragments prevents water from invading the conduit. Without water interference, rising basaltic magma then gives birth to the cinders and lava-flows typical of Strombolian eruptions. Thus, it was only the earlier parts of the eruptions of Capelinhos and Surtsey that were truly Surtseyan; their later stages comprised largely Strombolian eruptions, to which both volcanoes owe their armour-plating of lava-flows.

Surtsey must have been one of the most closely observed volcanic babies in history, for it could be relied upon to produce spectacular explosions that were far less dangerous than they looked. Surtsey grew from the ocean floor at a depth of 130m and emerged on 14 November 1963. Within four hours the eruption had developed its full fury, and had built up a cone 174m high when activity abated on 1 February 1964. The baton was taken up by a new vent that constructed a cone as big as the first in five weeks. The tuffs accumulated so quickly that water had increasing difficulty in reaching the vent. The Surtseyan eruptions were doomed and their fate was sealed on 4 April 1964, when lava-flows from Strombolian eruptions were emitted, which have helped protect Surtsey from wave attack.

The shores of many Atlantic islands are scattered with Surtseyan cones and prominent examples include Monte Brasil in Terceira and Monte Guia in Faial in the Azores, and El Golfo in Lanzarote and Taco in Tenerife in the Canary Islands. Surtseyan features also formed in ephemeral lakes during the pluvial periods in the deserts that corresponded to the ice ages in higher latitudes.

Icelandic eruptions

The hallmark of the Icelandic type of eruption is the emission of hot, fluid basalt through long fissures formed by rifting of the Earth's crust. Such eruptions are common in the broad swathe, or Axial Zone, of contemporary and recent volcanic activity, about 70km wide and 400km long, cutting diagonally across Iceland, which represents a rare instance of a section of a mid-ocean ridge built above sea level (Fig. 5.4).

In Iceland, the active zone is made up of a series of volcanic systems centred on an elliptical main volcano, from which emerge swarms of fissures, between 40km and 100km long, trending parallel to the direction of the zone as a whole (Fig. 5.5).

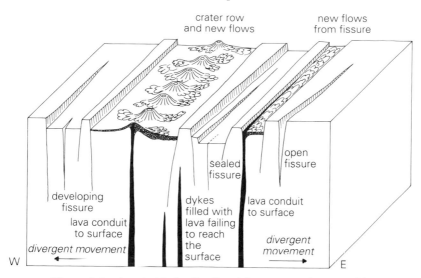

Figure 5.4 A typical Icelandic rift and fissure swarm (not to scale).

The fissures are caused by the divergence of the North American and Eurasian Plates, which stretches the crust for some 100–150 years before rifting supervenes. Only a small section rifts, however, at any one time. Since the settlement of Iceland in AD 874, sections of the active zone in northern Iceland have rifted and formed fissures in 1618, 1724–9, 1874–5, and in 1975–84. However, not all the fissures gave rise to surface eruptions, although magma almost certainly invades them at depth.

An Icelandic fissure eruption is beautiful, usually brief, and rarely dangerous. It

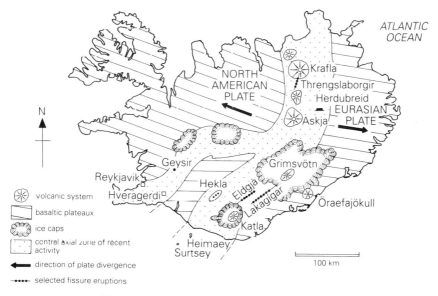

Figure 5.5 Selected volcanic features of Iceland.

begins with lava-fountaining, throwing many clots of hot lava, at temperatures of about 1200°C, high into the air like ornamental fountains of bright red paint. The emission is accompanied by a distinctive, sustained swishing noise as the associated gases escape. The clots weld together on landing to form spatter ramparts. After about two days, activity concentrates on a few vents, amassing cones of spatter and then cinders. The cones are then often breached by copious effusions of lava that soon spread out in thin, but extensive flows. Emissions reach an early crescendo with initial rates of effusion sometimes exceeding $1000\,m^3$ per second, but they soon become weaker and more intermittent, separated by lulls of increasing length until they stop almost unnoticed. The fissure then becomes clogged and sealed with solidified basalt, and rarely resumes its activity. The old fissure becomes a band of strength, perhaps 2m across and more than 20km long, that is added to the increasing width of the country. Thus, each newly formed fissure is paralleled by a multitude of lava-filled dykes. The typical result of Icelandic activity is a long line of low, broad cones, fringed by an apron of lava-flows, well illustrated by the Threngslaborgir, which erupted 2000 years ago in northeast Iceland. The most prolific basaltic eruptions in Iceland in historic time occurred on two fissures, which run parallel to each other, 5km apart, although they belong to different volcanic systems (Fig. 5.5).

The Eldgjà, "the fire-fissure", eruption took place in AD 934 on a fissure 57km in length, extending from the Katla system. After rifting, activity began with vigorous lava-fountaining, which was quickly succeeded by concentration on individual vents that gave rise to cones of cinders and spatter, and about $14\,km^3$ of fluid basalts were discharged that rapidly covered an area of $780\,km^2$.

The most famous of all Icelandic fissure eruptions, however, was the Lakagígar, or "Laki fissure" eruption on Laki mountain, extending 100km southwestwards from the volcano of Grimsvötn. After earthquakes had shaken out a warning for a week, the fissure opened on 8 June 1783 on the southwestern flanks of Laki mountain. Basalts gushed forth into the Skáfta valley and formed a thick and widespread lava-flow stretching more than 60km from the fissure and with an estimated volume of $10\,km^3$. The initial discharge of these very fluid lavas was very high, and an average of $5000\,m^3$ of lava per second was expelled – double the discharge of the River Rhine in Holland – throughout the first 50 days of the eruption. This southwestern fissure stopped operating when its northeastern extension suddenly opened up on the opposite flank of Laki on 29 July 1783 and directed great volumes of lava into the Hverfisfljot valley farther north. The flows were accompanied by more explosive activity that formed no less than 115 cones, usually ranging from 40m to 70m high, along the total fissure length of 25km. They were chiefly spatter cones and cinder cones although two tuff cones were also formed when water invaded part of the fissure. The lava-flows cover a total area of $565\,km^2$ and attain $12.3\,km^3$ in volume. Vast areas were blanketed with fine ash and two churches and perhaps 40 farmsteads were damaged. But the disaster which made the eruption notorious sprang from the copious emissions of poisonous gases, notably sulphur dioxide and carbon dioxide, that contaminated the pastures far from the fissure. As a result, more than three-quarters of the sheep and horses and half the cattle in Iceland died. The dreadful "haze famine" that ensued caused the death of over 9000 people (a quarter of the country's population), and many survivors, already living in poverty, emigrated. It was the greatest calamity ever to affect Iceland. Fine ash also fell in Scandinavia and Scotland, and a curious dry blue fog spread over Europe that Benjamin Franklin noted in his diary in Paris. The following winter was unusually cold in Europe.

Fissure eruptions have also completely dominated the formation of the island of São Jorge in the Azores. It is 54km long and rises more than 2000m from the floor of the Atlantic Ocean to a height of 1053m above sea level. Its crest is marked by small cones about 100m in height, but the bulk of the island is composed of basaltic lava-flows emitted from closely packed en échêlon bands of fissures. Since the Portuguese settlement in the 15th century, two small cinder cones were formed in 1580, and more than a dozen cones and basaltic flows were erupted in July 1808.

Hawaiian eruptions

Although fissure eruptions also occur on the Hawaiian volcanoes, they are really dominated by effusive lava emissions from vents clustered near a central spot (Fig. 5.1). The Hawaiian islands began life as seamounts, but eruptions have built them 4000m above sea level. Hawaiian eruptions produce hot, very fluid basalts with few and rather weak explosions. They accumulate in vast, gentle shields with slopes of 3–5°, culminating in a central hub that has collapsed in a shallow caldera, pockmarked by pit craters that sometimes contain molten lava lakes, such as Kilauea Iki which formed in 1959.

Most frequently, Hawaiian eruptions begin with lava-fountaining which spreads downwards along fissures as the gas is expelled. Sometimes small cones of fragments are formed as they did at Pu'u O'o in 1984, but calmer effusions usually follow, which may send long lava tongues down valley.

The Hawaiian volcanoes are related to a mantle hotspot, and it is above similar oceanic hotspots that comparable landforms have been developed in Réunion in the Indian Ocean and in the Galápagos Islands in the eastern Pacific Ocean. The Hawaiian volcanoes are the latest to be created by eruptions that have formed the submerged Hawaiian and Emperor chains of seamounts. The oldest eruptions took place over the Hawaiian hotspot some 75 million years ago, and the first volcanoes that they formed have been carried northwestwards on the Pacific Plate, and are now about to be subducted into the Aleutian Trench. In the Hawaiian Islands, Kauai is the oldest, between 3.8 and 5.6 million years old, whereas the youngest, Hawaii itself, emerged only about 700 000 years ago, and is, in fact, composed of five separate shield volcanoes. It is no doubt the northwestward movement of the Pacific Plate over the hotspot that explains why Kohala and Mauna Kea, the northwestern volcanoes on the island, are probably extinct (although Mauna Kea may only be dormant), whereas the present activity is concentrated on the southeast, in Kilauea and Mauna Loa and in the active seamount Loihi that still lies 1000m below sea level. Each volcano seems to have a life-span of about a million years. Kilauea, which first erupted above sea level about 100000 years ago, and Mauna Loa, which emerged about 500000 years ago and has accumulated a volume of 40000km³ of basalt, may be expected to make further contributions to the bulk of Hawaii before they retire into repose and are carried northwestwards beneath the Pacific Ocean like their predecessors. They still erupt frequently: Mauna Loa has had 39 eruptions since 1830 and Kilauea no less than 60. Magma also often rises into dykes and thus adds to the bulk of the island without actually reaching the surface. From 1979 to 1983, for instance, four eruptions took place on Kilauea, but 13 dyke intrusions failed to reach its surface.

Réunion Island marks the most recent of a series of volcanoes formed as the African Plate moved over a hotspot at a speed of about 3 cm a year (Fig. 5.6). Their remnants are found in the submerged Mascarene Plateau, 750km from Réunion, erupted about 40 million years ago; the now much-eroded island of Mauritius, 200km from Réunion, developed between 28 million and 18 million years ago; Réunion itself began to form on the floor of the Indian Ocean about 5 million years ago. Réunion is a massive accumulation of basaltic lava-flows that have formed two shields: one probably extinct, crowned by Piton des Neiges (3069m), and the other,

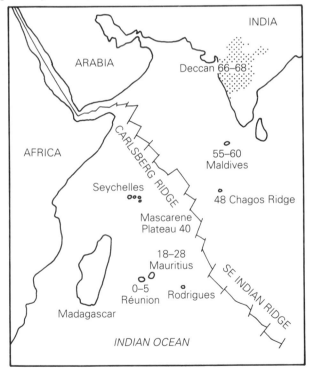

Figure 5.6 The hotspot track from the Deccan Plateau of India to Réunion Island, with ages of volcanic activity in millions of years.

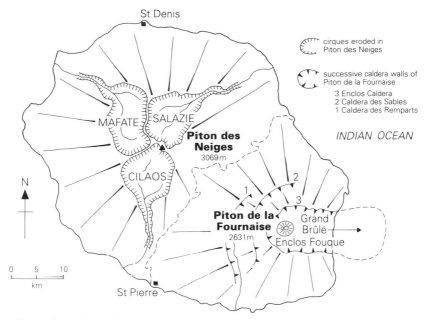

Figure 5.7 Piton des Neiges and Piton de la Fournaise volcanoes, Réunion Island.

Piton de la Fournaise ("Blazing Furnace"; 2632m), is amongst the world's most active volcanoes, erupting on average every ten months (Fig. 5.7). The base of the double volcano lies at a depth of 4000m and has a diameter of 220 km. The total volume of Réunion has been estimated to be at least 57000km³. Piton des Neiges emerged from the Indian Ocean about 2.1 million years ago, but probably became extinct some 12000 years ago. Piton de la Fournaise rose above sea level about 500000 years ago, and has been vigorously erupting ever since. During historic time, which in Réunion is about 400 years, three-quarters of the eruptions of Piton de la Fournaise have lasted less than a month and 39 per cent have lasted less than a week. On average, Piton de la Fournaise expels about 10 million m³ per year of fluid lava-flows discharged rapidly from the summit cone or its immediate flanks. About 95 per cent of the eruptions have been confined within the Enclos Fouqué, the youngest of three nested summit calderas. However, perhaps the most remarkable feature of the island is the Grand Brûlé, caused by vast landslides extending deep into the Indian Ocean.

Flood basalts

Eruptions of flood basalts may be caused when convection currents in the mantle generate wide plumes that arch up and fissure the crust. They have been well studied in the Columbia River and Snake River plateaux in Washington and Oregon in the United States (Fig. 5.1). Great volumes of basalt probably gushed from fissures, many of which were probably concentrated in the Grande Ronde volcano in north-east Oregon, which attained its vast volume and its modest peak about 12 million years ago. Many individual flows, exposed superbly in the Columbia River gorge, can be traced for over 150 km and cover more than 1000 km^2. The basaltic eruptions probably took place sporadically over millions of years, but were so fluid and so voluminous that they spread out easily over an area of 130 000 km^2. Surface slopes, of less than 2°, were so gentle that the volcano itself, if not its lavas, passed unnoticed until studies of its feeder vents in the Grande Ronde Canyon revealed its true significance. The Roza eruption marked one of its major effusions. Sheets of basalt were discharged at rates exceeding 1 million m^3 per second and covered 40 000 km^2 with 1500 km^3 of basalt in about a week. Although doubt has recently been cast on these figures, it has been calculated that 1 km^3 of lava was expelled from each 1000 m of fissure every day during the eruptions. At about the same time, a second centre in the Picture Gorge emitted the almost equally copious outpourings lying in the southern parts of the Columbia plateau. Similar emissions formed the plateau basalts of Antrim in Northern Ireland and of Skye in western Scotland about 60 million years ago. The largest basalt floods of the past, however, cover more than 500 000 km^2, for example in the Deccan plateau of India, in the Karroo basaltic area of South Africa, and in the Parana plateau in Brazil. The gaseous emanations that would almost certainly have accompanied such prolific eruptions, especially those of the Deccan plateau, have also been held responsible for the extinction of the dinosaurs. (It has been recently proposed that these plateaux were generated when large meteorite impacts cracked the crust and allowed great volumes of magma to escape to the Earth's surface. They could thus be the terrestrial equivalents of the lava expanses forming the lunar maria.)

It is the constant repetition of these voluminous eruptions of mid-ocean ridge, Icelandic, Hawaiian and flood basalt type that has helped make basalt the most common volcanic rock on the face of the Earth.

Strombolian eruptions

The Strombolian type of eruption is named after the activity on Stromboli in the Aeolian Islands, north of Sicily (Fig. 2.1). Strombolian eruptions usually expel basalts and sometimes andesites that form both cinder cones and lava-flows (Fig. 5.1). They are characterized by a moderate degree of lava fragmentation, moderate dispersal and moderate accumulations of both fragments and flows. The eruptions are short lived, lasting only a few months or years at the most; the explosions are all of similar intensity; and their cones rarely exceed 250 m in height or their lava-flows 10 km in length. The vent largely stays open during a Strombolian eruption. As the magma rises at an increasing pace into the upper reaches of the vent, more and more volatile bubbles separate out until it becomes frothy (Fig. 4.2). The bubbles burst and form an uprushing, expanding mass of steam and gas. The magma is shattered into innumerable clots and droplets and shot between 100 m and 1 km into the air during brief, raucous ejections repeated at intervals of perhaps an hour or less. The cold air solidifies these fragments into ash, lapilli, spatter or cinders as they return to the ground. The fragments are solid, if still glowing, when they fall back around the vent and quickly accumulate in a cone that approaches its final size within a few months. Each explosion produces fragments of a characteristic size, determined by its force, so that individual beds are therefore usually composed of similar materials. Few fragments are thrown far from the vent, although fine ash may be blown away on the winds. At intervals of several months, lava-flows ooze from the crater, or more often from the base of the cone.

At present, between four and six vents in the summit crater of Stromboli usually erupt every 15 or 20 minutes and lava-flows are commonly emitted several times a year and cascade down the Sciara del Fuoco, the great "scar of fire" on the northern flanks of the volcano (Fig. 5.8). The eruptions are usually too modest to make much impact on the media. Indeed, when Rossellini made the film "Stromboli" in 1950, he did more for the island's fame than any number of eruptions, although his interpretation of its eruptive style did not attain his usual realism. Nevertheless, these eruptions are the glory of Stromboli. What is already enthralling during the day becomes a beautiful pyrotechnic display when seen from the balcony provided by the Pizzo Sopra la Fossa, near the crest of the volcano (Fig. 1.2). The eruptions on the *piano nobile* below count amongst the most spectacular continuous natural performances on Earth. The first sign of another eruption is a low rumbling accompa-

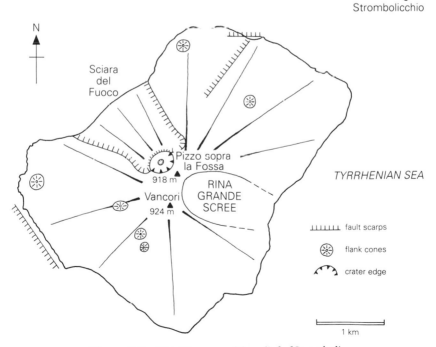

Figure 5.8 Chief features of the relief of Stromboli.

nied by a slight trembling of the ground. Then a dull red glow lights up first one and then another of the vents as the rumbling becomes a deep echoing roar of escaping gases. The red brightens inside the vents to pale vermilion, even yellow at times, as the molten rock arrives. One vent hisses out a blue flame like a huge Bunsen burner, while the roar in the other vents becomes raucous and hoarse. Incandescent fragments are hurled 100 m or more into the air, twisting and twirling in a constellation of all sizes from sparks of hot ash to clots of molten cinders, enveloped now and again by clouds of gas and steam. The fragments finally fall back gracefully onto the cones around the vents. The embers reveal the outlines of the cones for the first time as the glowing vermilion deepens to red again in the vent. As the sparks die, the larger clots slide slowly down the slopes of the pyre. The roaring in the vent calms to a subdued moan, and the noises made by the last cooled solid cinders crashing back down the vent echo about the crater. The silence returns as the last fragments on the cones cool and deepen to dark red before they finally vanish into the darkness. The whole sequence takes perhaps five minutes: there is less than half an hour to wait for the next.

A lucky observer may be rewarded with the eruption of a lava-flow. All the lava-flows in the past 400 years, at least, have been channelled down the Sciara del Fuoco. The molten stream cascades down the steep slopes and, as its surface solidifies, blocks often break off and roll quickly into the sea, generating explosions and billowing clouds of steam. The infrequent larger eruptions, when parts of the island may be showered with lapilli and even volcanic bombs, usually occur after periods of quiescence lasting perhaps several days. An abnormally vigorous eruption, albeit of limited danger, such as those occurring in 1930, 1956, 1967 and 1975, may then be confidently expected.

This kind of activity is perhaps the most common type on land. However, although Stromboli is typical of Strombolian eruptions, its shape and size are not. It is a strato-volcano, rising 3000 m from the sea-floor, which has been in continuous eruption for at least 2500 years. Such prolonged activity is exceptional, if not unique, amongst volcanoes and has created an unusually large landform as a result.

Parícutin is a well studied example of a Strombolian volcano that began erupting in a Mexican maize field on 20 February 1943 after a day of intense local earthquakes. Within a week, its cone was 167 m high. When, after a year, the cone reached a height of 336 m, Parícutin had already experienced most of its explosions and expelled most of its fragments. It repeatedly gave off lava-flows that changed gradually from basalts to andesites as the eruption progressed. They eventually covered 25 km^2 and overwhelmed the nearby town of San Juan. The activity slowly waned and then suddenly stopped on 4 March 1952 when the cone had reached the greater than average height of 412 m after an unusually long eruption. The Monti Rossi were more typical of the genre, erupting in three months on Etna in 1669. Once a Strombolian eruption stops, it often never resumes but it can, however, occur in distinct phases, forming superposed cones. Thus, in the Chain of Puys, the larger second cone of Puy de Mey completely buried the first. In continental environments, many Strombolian cones grow up along fissures, which can produce a succession of cones, erupted irregularly in time and space for several thousand years. These continental fissures thus behave differently from those in oceanic environments such as Iceland, where whole fissures seem to erupt only in one brief episode.

Apart from Stromboli itself, one of the best places to witness Strombolian eruptions in recent decades has been on Etna, especially since its contemporary phase of agitation began in 1971. It has erupted, for example, in 1974, 1978, 1981, 1984, 1985, 1986, 1989, 1990 and 1992. Etna has a collection of more than 200 Strombolian cones on its flanks as well as two large Strombolian cones, known as the North-East and South-East craters, near its summit. North-East crater started as a pit on 27 May 1911 at a height of 3100m at the base of the summit cone. After emitting gases and lava-flows for decades, it began to construct a cinder cone in 1955. Between 1966 and 1971, as the cone continued its growth, moderately viscous basaltic lava-flows were expelled which splayed out in a fan 4km wide and 200m thick. The northern sector of the cone collapsed under the combined influence of gravity and further basaltic upwelling in October 1974 and again in January 1978, so that it is now horseshoe shaped. Nevertheless, in January 1978 the cone was 250m high and formed, and still remains, the highest point of Etna at 3345m. Upon this achievement, the activity of North-East crater waned and only brief, sporadic outbursts have since occurred in 1980, 1981 and 1986. The baton was immediately taken up by South-East crater. It, too, had begun life as a pit, exploded out in 1971, but it only began to erupt lavas and form a cone on 29 April 1978. Its initial eruptions were punctuated by periods of repose, but it began persistent activity in 1980 that still continues – including, for example, two weeks of spectacular lava-fountaining in September 1989 rising 300m above the cone. South-East crater has expelled lava flows and built up a cone at a faster rate than its predecessor, and it may eventually take over the rôle of the summit cone on Etna.

Vulcanian eruptions

The Vulcanian type of eruption was designated from studies of the activity of Fossa cone at Vulcano in the Aeolian Islands between 2 August 1888 and 22 March 1890 (Fig. 2.1). It is a repetitive, noisy event, full of sound and fury, characterized by high fragmentation, with a quite wide dispersal away from the vent and fairly thick, fine accumulations (Fig. 5.1). Vulcanian activity gives off much gas and far more ash, cinders, and pumice than lava-flows from its rhyolitic, trachytic or andesitic magmas. Much ash and pumice is blown away from the vent, and the cones are small and squat. Like Fossa itself, they commonly rise between 350m and 600m high, with a diameter approaching 1km, and outer slopes of between 35° and 40° (Fig. 2.3). Vulcanian activity usually lasts, with long intervals of repose, for several millennia. It is similar to hydrovolcanic activity, but it is dominated by evolved magmas and its repose periods are much longer than those associated with basaltic magmas.

Vulcanian eruptions burst into life suddenly after a dormant period that can last for over a century. They begin with a violent, noisy, phreatic explosion, but soon afterwards shattered fragments of both new and old magma are expelled with loud bangs like gunfire or bomb detonations. It was no doubt such noises that encouraged the Greeks and Romans to believe that Vulcano was the site of the forges of Vulcan. Very dark grey, billowing ash clouds rise high into the air, and darken the sky. Near the vent, ash, dense lapilli, pumice, and breadcrust bombs, with a surface like a French loaf, build up a cone that usually retains a deep, clearly funnel-shaped crater. The eruptions continue for several months, with many active spasms lasting for a day or two between intervals of total calm. When the eruptions end, a dome may rise in the crater, but a lava-flow is equally likely to be extruded.

Fossa has been the main theatre of activity on Vulcano for the past 14000 years (Fig. 2.3), but in the 2000 years of historic time on the island, eruptions have averaged only about one a century. It is largely composed of fine ashy tuffs of trachytic composition, and has also produced two rhyolitic flows (the Commenda flow 1200 years ago and the Pietre Cotte in 1739) and two trachytic flows (the Palizzi flow about 1600 years ago and the Punta Nere flow about 5500 years ago). Fossa is crowned by the Gran Cratere, 500m in diameter and 175m deep, with near-vertical lower walls and splayed upper walls of about 35° (Fig. 2.4). Most of its products are trachytic, expelled in four main cycles of activity, each lasting several centuries. The bulk of the cone was formed in the first cycle ending about 5500 years ago. The present cycle, which began in 1444, has been characterized by smaller, rather more frequent eruptions than previously. The latest eruption produced only a maximum of 5m of fragments around the crater rim, and scarcely altered the shape of the crater itself, being probably the mildest eruption of Fossa ever. If apparent trends continue, however, another eruption seems likely to occur before the century ends.

The most diligent Vulcanian performer of the present century has been the Japanese volcano Sakurajima, which has experienced almost daily explosions since 1955. Another Japanese volcano, Asama, in 1783, and Arenal in Costa Rica, in 1968, also had major Vulcanian eruptions. On 19 February 1975, the Vulcanian eruption of Ngauruhoe in New Zealand lasted only five hours and was marked by explosions like the sound of gunfire at interludes ranging from 15 to 50 minutes. Equal proportions of old and fresh magma were ejected and most of the new magma was andesitic. Each explosion threw up a column to a height of some 4km, but many quickly collapsed and surged down the flanks of the cone.

Phreatic and phreatomagmatic eruptions

Phreatic explosions expel steam and shattered country rocks with no new magma; phreatomagmatic eruptions eject greater proportions of shattered new magma in addition to steam and country rocks, which still, however, often comprise some 90 per cent of the fragments. Both are characterized by violent steam explosions generated when water interacts with rising magma below ground. Many eruptions occurring on land have some phreatic or phreatomagmatic component because fissures in the Earth's crust allow surface water to seep down to meet the magma.

Such eruptions are usually brief because the water is limited in volume and restricted in supply, compared with the amounts available in Surtseyan activity. They are, however, often very violent because they take place in a confined space under ground. These eruptions form diatremes below ground and explosion craters, or maars on the surface, surrounded by a rim of shattered fragments consisting of varying proportions of new magma and country rocks.

The explosive vigour of virtually all types of magma is greatly increased by their interaction with water, which is suddenly changed to steam. But, proportionately, as in Surtseyan activity, the greatest increases in violence are added to basaltic magmas, which would usually otherwise only be mildly explosive.

Events begin with the encounter of two fluids at different temperatures at depths commonly exceeding 250 m. Fuel–coolant interactions take place. The coolant, the water, is at temperatures ranging, say, between 20°C and 100°C. The magma is quenched; heat is suddenly transferred from magma to water, which expands explosively into steam. The thermal energy of the magma is transformed into the considerable mechanical energy of the steam in a fraction of a second. The magma is shattered, increasing the area in contact with the water, thus accelerating steam formation that, in turn, causes more shattering. The rocks are very quickly ruptured, a trumpet-shaped conduit is formed and a circular explosion crater is blown out at the surface. Fine, angular fragments of country rock with some new magma are shot out of the vent. They form columns of ash and steam rising 5 km or more into the air, which sometimes collapse and radiate from the vent like the collars spreading from the base of a nuclear explosion. The explosions begin violently and often unpredictably, and reach their maximum power in seconds, but then wane rapidly so that each pulsation rarely lasts more than half an hour. A succession of blasts often occurs, separated by increasing intervals of repose, but none is as violent as the first.

Repeated explosions weaken the walls of the conduit, which collapse and slump downwards along circular faults that develop. Thus, the conduit is widened, but it is also often choked temporarily with rock debris. The blockages are, however, then removed by further explosions at greater and greater depths as more water continues to infiltrate. The whole episode only lasts a few hours, or a few days at the most – primarily because the restricted water supply is soon eliminated by the explosions. Thus, the eruption ends when the water supply is exhausted, and not, as in most other eruptions, when the magma stops rising. A collapsed and shattered mixture of magma and country rock fragments then chokes the lower parts of the conduit, often to depths exceeding 1000 m, which thus forms a diatreme piercing the crust. Coarse angular fragments of country rock and a small proportion of shattered new magma come to rest in a thin rim or crescent accumulated on a sector around the deep, circular explosion crater, which is commonly 1 km across. Rainwater quickly forms lakes in these craters, which are usually known as maars ("meers") from their name in the Eifel region in western Germany where they were first described. In that area, 72 maars were formed where streams filtered down to meet rising basalts, but, where water did not infiltrate, the same magma formed 160 Strombolian cones. The maars are often 500 m to 1 km across and up to 200 m deep: Pulvermaar, for instance, is 900m in diameter and Gemundenermaar is 204m deep. Most were formed between about 10 000 and 12 500 years ago, during the decline of the last ice age when abundant water became available when the permanently frozen ground began to melt. They are also common on similar volcanic plateaux in Auvergne, where over 20 maars have been identified in, for example, the Chain of Puys. Farther south, in Velay, lie the Lac du Bouchet, 800m across, 30m deep and enclosed in a wooded rim of fragments, and the Lac d'Issarlès which fills a fine crater 110m deep and 1km in diameter. Lake Nyos, one of 29 maars in the Grassfields region of western Cameroon, has become one of the best known maars in recent years – not, however, for the phreatomagmatic events which created it some 500 years ago, but for its notorious emission of carbon dioxide in 1986. Lake Nyos is 208m deep, and is larger than most maars, with a diameter of 1.8km from north to south and 1km from east to west. Its rim is marked by a crescent where basaltic fragments form thin layers between large shattered blocks of the granitic country rock.

Phreatic eruptions are frequent precursors of magmatic eruptions on many strato-volcanoes; and one of the most sustained periods of phreatic activity this century occurred on the dacitic dome of Lassen Peak in the Cascade Range. The first 170 explosions between 29 May 1914 and 14 May 1915 were phreatic, expelling steam and fragments shattered from the dome that sometimes rose 3km above the summit. The main explosions occurred at quite regular intervals: 14 June, 18 July, 19 August, 21 September, 22 October . . . perhaps representing the time required for snowmelt and rainwater to infiltrate the dome and meet the magma. At first, the fragments ejected were not even warm enough to melt the summit snows; and the first hot ash did not emerge until 22 October 1914, showing that magma was approaching the surface. On 14 May 1915, rapid snowmelt caused the last phreatic throat-clearing explosion that opened the vent for the first emission of magma and a change of eruptive style.

Phreatic explosions were expressed differently in September 1979 on Etna, whose summit was then marked by two deep craters, the Voragine and the Bocca Nuova (Fig. 5.9). On 2 September 1979 part of the damp walls of the Bocca Nuova collapsed, thereby blocking the emission of steam and gases from that vent. With the temperatures of 1000°C prevailing at depth, the collapsed material was soon heated to incandescence and its humidity was converted to steam that joined the gases unable to escape. Great pressure consequently built up until it was released by a sudden explosion which occurred without warning at 17.15 on 12 September. The blocking rocks were shattered and thrown from the vent, killing nine tourists and injuring many others as they fled. More than a hundred comparable eruptions, fortunately without fatalities, have apparently occurred on Etna during the past century.

A less violent form of phreatomagmatic activity occurs when molten lava-flows invade marshy ground, or a shallow lake where the lavas are thicker than the depth of the water. The waters are trapped beneath the lavas and suddenly heated to steam, which then explodes through the flow. As a result, some of the lavas shatter into fine tuffs, forming gently sloping low cones between 10m to 50m high, although a few rare examples reach 100m. They are called rootless cones because their vents do not

Figure 5.9   Bocca Nuova, Etna.

extend below the base of the flow. South of the Chain of Puys, the lava-flow that erupted about 13 700 years ago from Puy de Tartaret invaded the valley of the River Couze Chambon and formed about 30 rootless cones. There are perhaps 10 000 rootless cones in Iceland where abundant surface water and frequent lava eruptions provide necessary ingredients. One of the finest series was formed in Lake Myvatn when the Younger Laxa lava-flow erupted from the Threngslaborgir fissure and invaded its shallow waters about 2000 years ago.

Subglacial eruptions

Depending on its situation and thickness, an ice-cap adopts two contrasting rôles in relation to volcanic activity. When an eruption occurs beneath a thin cap on a large volcano, lahars are often generated. These are given separate treatment later in this chapter. Eruptions beneath a thick ice-sheet are of a different style, dominated by melting ice and deep meltwater, where the spread of the erupted materials is inhibited at first by the still unmelted surrounding ice (Fig. 5.10).

Eruptions beneath a widespread ice-sheet exceeding 1 km in thickness have occurred in Iceland since the beginning of the ice ages. The eruptions themselves were typically Icelandic basaltic emissions, but the volume and pressure of the overlying ice sheets served at first to limit the extent and power of the eruption. However, it has been estimated that molten basalt erupted beneath the ice-sheets could melt ten times its volume of ice, and that, given the rapid effusion rates usually pre-

51

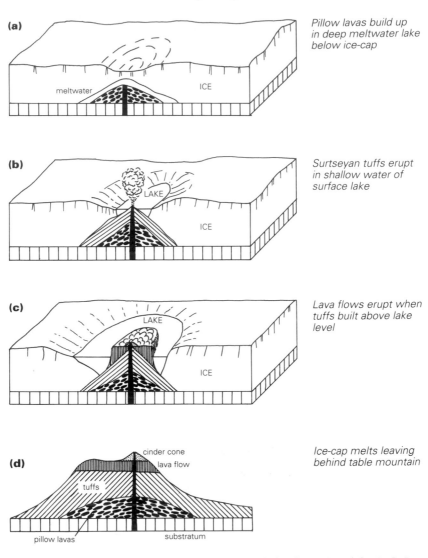

(a) Pillow lavas build up
in deep meltwater lake
below ice-cap

meltwater   ICE

(b) Surtseyan tuffs erupt
in shallow water of
surface lake

LAKE

ICE

(c) Lava flows erupt when
tuffs built above lake
level

LAKE

ICE

(d) Ice-cap melts leaving
behind table mountain

cinder cone
lava flow
tuffs
pillow lavas   substratum

Figure 5.10   The formation of table mountains by subglacial eruptions (after J. G. Jones, *Geological Society of London, Quarterly Journal* **124**, fig. 2, p.202, with the permission of the Society).

vailing during the initial stages of Icelandic eruptions, about 1 km³ of meltwater could be formed in less than a week. As a result, a subglacial water-filled vault or lake rapidly developed over the vents, but its waters were often prevented from escaping by the enveloping ice. The basalts thus erupted into the deep waters as pillow lavas, which piled up in a steep-sided mound as the eruption continued, because the surrounding ice also prevented the lavas from spreading sideways. Meanwhile, escaping hot gases also melted their way to the surface of the ice-sheet as the hot lavas accumulated in the meltwater vault below. Hence the ice-cap was eventually pierced and a lake, surrounded by the unmelted ice-sheet, was exposed at the surface. Pressures on the lavas were then much diminished, and the eruption changed in character although the nature of the magma did not alter. The formation of pillow lavas ceased and was replaced by the explosive interaction of rising magma and shallow lake water, producing typically Surtseyan eruptions, that built a squat cone of fine, glassy fragments above lake level. Because they are only active for a brief period, many fissure eruptions never got beyond this stage before extinction supervened.

However, where larger central volcanoes were active longer, continued ascent of magma eventually sealed meltwater access to the vent, so that basaltic flows could emerge and formed a lava-cap protecting the Surtseyan tuffs and the pillow lavas beneath. When climatic change caused the ice-sheets to melt, these accumulations were revealed as isolated, steep-sided "table mountains" (or "tuyas", as they are known in British Columbia). The most famous example in Iceland is Herdubreid, which is 1000m high and has a basaltic cap some 2km in diameter.

Subglacial eruptions in Iceland have also been responsible for jökulhlaups, or glacier-bursts. They are generated when an eruption melts ice and the meltwaters escape down slope. The escape occurs at speed and jökulhlaups are impressive floods lasting several days, with discharges approaching that of the River Amazon. In March 1972, for example, the jökulhlaup generated by an eruption of Grimsvötn discharged about 2km^3 of water in three days. They form the sandy plains fringing many Icelandic mountains, such as that at the foot of Öraefajökull, which developed a notable jökulhlaup in 1362. The main culprit in contemporary Iceland is Grimsvötn, which regularly erupts beneath Vatnajökull, the largest ice-cap in Europe.

Pelean eruptions and nuées ardentes

When Montagne Pelée annihilated St Pierre on Ascension Day 1902, a previously unrecognized kind of eruption was thrust upon the scientific world. Pelean eruptions are characterized by high degrees of fragmentation and thick accumulation, but dispersal is limited to about 50km^2. They develop chiefly on strato-volcanoes mainly from rhyolitic and dacitic, but also from trachytic and andesitic magmas (Fig. 5.1). What sets Pelean eruptions apart from others is the dominant combination of dome-formation, nuées ardentes, and laterally directed blasts, which are very destructive because they often concentrate their energy at ground level.

Pelean eruptions are separated by decades when no activity is displayed. Warning ejections of ash begin a few weeks before the Pelean culmination and continue in spasms of decreasing intensity for several months afterwards. The climax develops very suddenly when nuées ardentes at 700°C or more blast from the volcano. Subsequently, the viscous domes that often rise out of the vents may be partly destroyed by further eruptions. As the eruptions become weaker, the domes solidify, strengthen and eventually survive as solid corks capping the vent until the next eruption. The dome at the summit of Lassen Peak has an exceptional height of 500m, which, no doubt, enabled it to resist the small eruption of 1915 almost intact.

53

Nuées ardentes have captured the imagination ever since St Pierre was destroyed. If purely aesthetic considerations could be divorced from their effects, then nuées ardentes would be as attractive as their name suggests, but they are often the greatest killers in the volcanic repertoire. The direct translation of nuée ardente is "incandescent cloud", but "glowing avalanche" is, perhaps, a more graphic and accurate description. Close investigations have shown that they encompass a continuum of variations, and many different terms have been used to describe them. They are, in fact, prominent features of both Pelean and Plinian eruptions. Nuées are a type of ashflow, propelled down slope in avalanches of a hot, frothy, dull-red, aerosol-like emulsion of gas, fluid lava, old blocks, and densely packed fragments of molten and chilled ash and pumice. They are commonly accompanied or preceded by similar, but less dense, ground surges and hot blasts, propagated at supersonic speeds and frequently spreading 10 km from the vent. In Pelean eruptions, many nuées ardentes may succeed each other during several months, although the first, like the one that destroyed St Pierre, is usually the most powerful. Nuées ardentes move fast. Billowing clouds of ash and gas rise above them as they roll over the ground in enormous bursts at speeds of up to 500 km an hour. Photographs often give a misleading impression of nuées ardentes because they usually reveal only the billowing clouds and mask the damage-dealing, thin, surging avalanche below them. They are brief, lasting only a few minutes from initial blast to smoking wreckage.

By coincidence, a different nuée ardente had erupted from the Soufrière of St Vincent on 7 May 1902, the day before St Pierre was destroyed, but it only killed 1565 people and was immediately overshadowed by its rival. The St Vincent nuée was first directed vertically, perhaps because it was funnelled upwards by a deep confining crater, in an eruption that was more violent, in fact, than that at Montagne Pelée (Fig. 5.11). It rose in a column high into the air, cooled, waned and collapsed, and then spread in a radiating base-surge all around the flanks of the Soufrière. This type of nuée ardente seems to be the most common generated in Plinian eruptions. Because the St Vincent type of nuée ardente travels up through the air, its fragments are often sorted into layers of similar size when they fall back to Earth, unlike the unsorted mixtures typically thrown out by Montagne Pelée. The base-surges spread outwards in hot gusts, forming hummocky, dune-like landscapes. Not to be outdone, Montagne Pelée itself produced a St Vincent type of nuée ardente on 30 August 1902 that devastated an area of 114 km^2, and killed about 1000 people who had hoped they were safe on the eastern slopes of the volcano.

Most nuées ardentes are erupted only infrequently, but Merapi in Java regularly emits a third, less explosive and less dangerous type of nuée ardente (Fig. 5.11). Nuées are continually released from the base of the dome growing at the summit of Merapi. As it swells, the dome bulges over the western flank of the volcano like a soufflé, and finding nothing to rest on, its outer shell collapses and crumbles down slope, causing the sudden relief of pressure that releases the nuées ardentes. It is the topographic position of the Merapi dome that determines the frequency of its eruptions. If the dome were supported, it would probably grow larger and would only be destroyed by more violent, but less frequent, truly Pelean nuées ardentes.

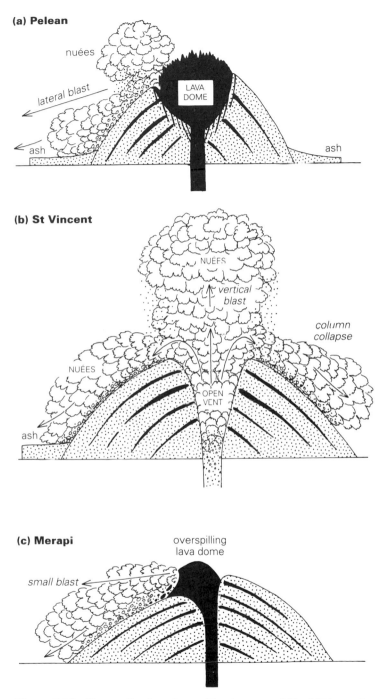

Figure 5.11 Types of nuée ardente. (Based on page 149 of *Volcanoes* by Gordon A. Macdonald, ©1972. Reprinted by permission of Prentice–Hall Inc., Englewood Cliffs, NJ 07632)

Thus, the most dangerous types of nuée ardente are influenced by the relief of the volcano. The paths that they take are dominated by gravity. They hug the ground. That is why they can kill so many people so quickly.

Plinian eruptions

Plinian eruptions include the most violent forms of all volcanic activity (Fig. 5.1). They are the eruptions of all the superlatives and the attempt to describe them stretches the vocabulary of a language even as rich as English. Their name commemorates Pliny the Elder, the famous victim of the classic eruption of Vesuvius, and his nephew, Pliny the Younger. Plinian eruptions are characterized by intense fragmentation of the magma, its enormous dispersal, commonly over more than 500km^2, and by the thick accumulation of materials that can exceed 100km^3 in volume. They unleash some of the most exceptional and terrifying power in the natural world, releasing megatonnes of energy and considerable volumes of magma at great speed. It was reported, for example, during the great eruption of Cosegüina, in Nicaragua in 1835, "that wild beasts, howling, left their caves, and snakes and leopards fled for shelter to the abodes of men". And humanity is not immune: the terrifying Plinian eruptions of Vesuvius have induced many a Neapolitan sinner to rush to confession. Some of their more tangible effects are recounted in the later descriptions of the eruptions of Vesuvius, El Chichón, Mount St Helens and Pinatubo. Only a few Plinian eruptions occur in a century and, because they often start suddenly and unexpectedly after a long quiescent period, the chances of a trained observer being on the scene when they begin are remote. It is also hard to observe them closely without being killed.

These outbursts expel fine ash and pumice, composed of rhyolite, dacite, phonolite, trachyte or andesite often at supersonic speeds. Ashfalls, ashflows and nuées ardentes predominate. Lava-flows are only emitted as the eruption ends, if at all. Plinian eruptions very rapidly develop a towering column of gas and ash, sustained by discharging, exploding volatiles and convection, which spreads out like the branches of a pine tree in billowing, cauliflower-shaped clouds. This commotion often causes violent thunderstorms, whereupon the rain welds the ash into bullet-sized lapilli. Ash may spread 200km from the volcano, whereas the finest dust, carried by the winds and high stratospheric currents, can circle the globe. The Plinian eruption of Laacher See in western Germany 11 000 years ago expelled a layer of ash that has been recognized over much of central Europe. Ash resembling fine flour from Mount St Helens choked car radiators, and had to be shovelled like snow from the streets in eastern Washington within hours of the eruption on 18 May 1980, and the dust expelled from Krakatau gave brilliant sunsets throughout the world after the Plinian eruption in 1883. Coarser ash, lapilli and, especially, pumice fall nearer the vent, and rained down from Vesuvius upon the hapless people of Pompeii in AD 79. Total and terrifying darkness may envelop the region for many hours, as witnessed on the coastal areas of Sumatra and Java facing Krakatau in 1883.

If the vent is widened, or if the volatile content of the eruption is reduced, the column collapses, and surges down the volcano as nuées ardentes, destroying anything on the surface in a searing heat such as that which sealed the fate of the villagers on the slopes of El Chichón in 1982 and the Pompeiians in AD 79. Lahars then also often develop, as occurred at Mount St Helens in 1980 as well as at Pinatubo in 1991. As the eruption ends, the whole summit area may collapse into a caldera in a Wagnerian climax. The crests of Pinatubo in 1991 and of Coseguina in 1835 were both partially destroyed by such outbursts. A Plinian eruption builds no cones: it is a cataclysm that is far more destructive than constructive. But, although a Plinian eruption commonly leaves a decapitated volcano with a gaping, devastated hole in its summit, a fuming dome may soon begin to rise within it that heralds the renewal of construction.

The average Plinian climax lasts for less than two days, and nowhere in nature are such vast changes wrought in such a short time. The energy discharged is many times greater than in the most powerful bomb yet exploded. At Vesuvius, the classic Plinian eruption was succeeded by many others, notably in 472 and 1631. Similar outbursts have made a multitude of volcanoes infamous, such as Tambora in 1815, Coseguina in 1835, Krakatau in 1883, Santa Maria in 1902, Bezymianny in 1956, Mount St Helens in 1980, and Pinatubo in 1991.

It is Plinian eruptions that usually expel volcanic dust and gas into the stratosphere, and form volcanic aerosols. In June 1991, for example, dust clouds from Pinatubo proved a distinct aviation hazard. In the month following the eruption on 15 June 1991, at least 14 airliners flew into dust clouds and nine were obliged to make emergency landings when the particles severely damaged their jet engines. Neither radar nor the human eye can effectively distinguish between such dangerous clouds and harmless normal clouds at high altitudes. However, as a result of the first international conference on volcanic ash and airline security held in Seattle in July 1991, communication will, at least, be improved between volcanological observatories and air-traffic controllers, so that pilots may soon have advance warning of dust clouds along their flight paths. Fortunately, the bulk of the ash and dust from Plinian eruptions falls back to Earth after several months.

The formation of volcanic aerosols has more extensive effects. These aerosols eventually envelop the whole Earth, persisting in the stratosphere, usually at an altitude of between 15 km and 30 km, for several years. Pinatubo is only the latest in a long line of quite reliably documented culprits, stretching back to the eruption of Tambora in 1815. Pinatubo expelled fluorine and chlorine compounds, but chiefly sulphur dioxide, into the stratosphere. The sulphur dioxide is changed into sulphur trioxide, which then combines with water to form droplets of sulphuric acid. These droplets reduce temperatures by reflecting part of the solar radiation back into space, thus diminishing the amount that can reach and warm the Earth's surface. Their most marked cooling effects seem to be experienced in northern North America and northwestern Europe, mainly for between 6 and 18 months after the Plinian eruption. The aerosols from Pinatubo encircled the intertropical zone, covering 40 per cent of the Earth's surface within a month of the eruption in June 1991, and encompassed the globe before the end of that year. Their volume was twice as large as that expelled by El Chichón in 1982, although Tambora probably ejected at least five or six times as much in 1815. If these estimates can be extrapolated into climatic effects, it seems that Pinatubo will have a greater influence on world climate between 1991 and 1995 than El Chichón had after 1982, but significantly less than Tambora accomplished, especially in 1816.

Nevertheless, lest all Plinian eruptions be blamed for poor weather, it should be emphasized that only those outbursts giving off vast volumes of sulphur dioxide, possibly in association with carbon dioxide, anhydrite, fluorine or chlorine, as well as unusually large quantities of dust, have appreciable climatic consequences. In 1980, for example, Mount St Helens had only a minimal and localized effect on the weather, compared with El Chichón, which erupted on a similar scale. Moreover, very large-scale basaltic emissions, such as that of Lakagígar in 1783, have also been associated with sulphur dioxide and carbon dioxide clouds.

The Plinian eruption of Tambora on Sumbawa island in Indonesia was the largest eruption of historic time. It took place on 10 and 11 April 1815, and the explosions were heard more than 500km away in Java. Its eruptive column rose 40km high, and Sumbawa and nearby Lombok were totally dark for two days and were blanketed with about 50cm or more of ash and pumice that killed trees, buried crops, collapsed roofs, and killed an estimated 10000 people. Ephemeral floating islands of pumice clogged the sea for weeks afterwards. The terrified survivors of the cataclysm then suffered starvation and often death in the ensuing famine and 92000 people are thought to have died as a result of this eruption – the largest volcanically related death-toll in historic time. Before the eruption, Tambora rose to 4000m, but the outburst decapitated its summit by 1000m and formed a deep caldera, over 6km in diameter, whose floor lay 700m below its jagged rim. At least 40km^3 of fragments were thrown into the air, and perhaps even double that amount was expelled as nuées ardentes and ashflows. The finer dust particles and sulphur aerosols circled the Earth for several years and have been held responsible not only for a series of brilliant sunsets that inspired Turner, but also for "the year without a summer" in Europe and North America in 1816. Frosts were frequent in June in New England; European wine harvests in 1816 were the latest since records were first taken in 1482; and summer temperatures in northern Canada were 3°C or 5°C below normal. Brilliant sunsets notwithstanding, it is fortunate that Plinian eruptions are so rare.

The blankets of ash that cause such devastation, however, have geological uses, as well as adding valuable nutrients to the soil. If old ash layers can be identified and dated accurately, they provide invaluable indicator beds that are as useful in marking geological time as a clock striking in the night. In this respect, the ash scattered by the Plinian eruptions of Hekla have been of paramount importance in establishing the volcanic chronology of Iceland. Thus, the nastiest volcano in the country has been one of the most useful from a geological point of view. Five major eruptive cycles have been identified, each with a similar pattern. They began about 6600 years ago (Hekla 5), about 4000 years ago (Hekla 4), about 2800 years ago (Hekla 3), and between 1500 and 2000 years ago (Hekla 2); the latest cycle (Hekla 1) began in AD 1104. Plinian eruptions inaugurated each cycle and mantled much of Iceland with identifiable rhyolitic ash that contrasts well with the basalts prevailing over most of the country. In the latest cycle, until 1947 the main outbursts occurred at intervals of about a century. The inaugural Plinian eruption in the autumn of 1104 spread fine rhyolitic pumice northwards over half the country and caused many farms to be abandoned. Other notable Plinian eruptions occurred in July 1300 and on 25 July 1510, when volcanic bombs, almost 1 m across, were thrown 40 km from the vent. The pumice blanket expelled on 13 January 1693 was accompanied by poisonous gases that killed many animals, and damaging ash was also expelled in April 1766 and 2 September 1845. The Plinian eruption on 29 March 1947 was one of the most powerful of the present cycle and 1 km^3 of materials were ejected in 13 months. The Plinian column rose 27 km in 20 minutes when the eruption started. Pale rhyolitic ash and then brown dacitic ash soon covered 70 000 km^2, and within two days dust was falling in Finland – providing an indicator bed for future geologists. Something, however, has been disturbing Hekla in recent decades. It erupted on 5 May 1970 and then again on 17 August 1980, when it expelled a Plinian column, 15 km high, that spread fluorine-rich rhyolitic ash over about 17 000 km^2. A short-lived modest ash column, 6 km high, was formed during another eruption that lasted only a week from 9 April 1981. Hekla erupted yet again on 17 January 1991 and formed a Plinian column that rose to a height of 11.5 km in ten minutes. But this brief outburst delivered only some 0.02 km^3 of fragments, the smallest volume Hekla has expelled in the present cycle.

Flows and floods of pumice and ash

Flows and floods of pumice and ash are very similar to nuées ardentes; they form just as rapidly, but cover much larger areas and, although several such eruptions have taken place in the past few million years, only one, at Katmaï in Alaska in 1912, has been recorded in historic time. Characterized by high fragmentation and great thickness, their dispersal is often limited by relief features. Huge volumes of fine, frothy pumice emerge almost without explosion and spread outwards rather like milk boiling over from a pan. The chief difference between flows and floods is size. Both are composed of more evolved magmas: rhyolites are often associated with the more voluminous eruptions, but trachytes and dacites have also been expelled.

About 21 000 years ago, the Ito ashflow from Aira volcano produced 150 km^3 of pumice in one of Japan's biggest eruptions. In central Oregon, the rhyolitic ashflows of the vast John Day Formation, which are well displayed in the "Flood of Fire" trail near Mitchell, gushed out about 20 million years ago, killing a multitude of plants and animals that now make it an invaluable source of fossils (Fig. 5.12).

Rhyolitic floods are on an even grander scale. It has been calculated, for example, that the Taupo eruption in New Zealand covered an area of 20 000 km^2, and the rhyolitic floods underlying much of Yellowstone Park in Wyoming have a volume of 3000 km^3. Some rhyolitic floods were probably expelled at speeds in excess of

700 km an hour. If they emerge at temperatures over 600°C, the fine ash and pumice can weld together into massive, tough accumulations known as ignimbrites, which often look like lava-flows. The scale of rhyolitic floods was only truly appreciated after the eruption of Katmaï on 6 June 1912. This was one of the largest eruptions of the century and no less than 35 km³ of pumice frothed out from near the foot of Katmaï, swamping the Ukak valley to a depth of 200 m within 60 hours. Decades later, it was still emitting steam and gas that merited the name "Valley of Ten Thousand Smokes", given by those lucky enough to witness it. Now the smokes are extinguished and all that remains is a thick layer of welded, unsorted sandy pumice.

The eruptions of Mont-Dore volcano in Auvergne reached their climax about 3 million years ago when a vast rhyolitic flood known as the "Grande Nappe" was emitted from fissures on its northern flanks. In the space of perhaps less than a day, some 10 km³ of yellow ash spread over 360 km². It was remarkably friable, however, and only survives where it has been protected by later lava-flows.

The major eruptions in the Phlegraean Fields, west of Naples, apparently occupy an intermediate position between flows and floods. Between 36000 and 27000 years ago, a series of large eruptions formed the Campanian Ignimbrite of welded fine trachytic fragments which covered an area of 7000 km² and had a total volume of about 80 km³. A smaller eruption, in a brief episode some 12000 years ago, formed the trachytic Yellow Neapolitan Tuff, which covers much the same area. It is not yet clear whether either formation was emitted as a single unit or in a series of eruptions, and thus whether they constitute a flood or an accumulation of smaller ashflows.

Figure 5.12   John Day Formation ashflows, "Flood of Fire Trail", nr Mitchell, Oregon.

Blasts and debris avalanches

The special effects of blasts and their associated debris avalanches were only fully realized after Mount St Helens erupted in 1980 (Fig. 5.1). Since 1980, more than 150 debris avalanches have already been identified which formed during the ice ages and at least 17 that have occurred in the past 400 years. They are characterized by low fragmentation, thick accumulations, and moderate dispersal. The debris avalanche does not, in itself, constitute a volcanic eruption. It is, however, precipitated when magma rises asymmetrically into the cone, making one sector of the volcano bulge outwards and become unstable. This sector can then fail in response to the earthquakes, or to the pressures caused by rising magma. The consequent collapse of the flank is sudden and decisive. The mountainside founders in a giant, rippling landslide or debris avalanche. The sudden reduction in pressure confining the magma unleashes the true eruptive phase. Hot gases in the magma blast out sideways in a major explosion that is immediately followed by nuées ardentes or ashflows. At Mount St Helens, the hot blast, at a temperature of 200°C, travelled faster than the debris avalanche and spread much farther. Trees and buildings were razed to the ground and scorched, especially on the sides facing the volcano. Although blasts devastate wide areas, their deposits are often so thin and fine that they had often been misinterpreted before 1980 – if they had been noticed at all. The blast deposits at Mount St Helens, for example, covered an area of 600 km^2, but averaged less than 10 cm in thickness even immediately after the eruption and before they could be eroded away by the rains.

The debris avalanche itself is a much more substantial feature, for it comprises the parts of the original volcano suddenly transferred down slope at speeds of 250 km an hour. At Mount St Helens, 3 km^3 of the volcano was spread over 60 km^2. Most debris avalanches can be traced back up slope to their place of origin on the flank of the volcano, which is marked by a horseshoe-shaped caldera, opening out where the debris left the flank. At the exit to the caldera, it sweeps a smooth path, but it soon becomes typically hummocky with hillocks commonly reaching 400 m long and 20 m high. Debris avalanches commonly reach 300 m in thickness and are composed of a mixture of fragments of all sizes, ranging from fine clays to lava blocks, some-times transported intact and reaching 100 m across, from the body of the volcano. Unlike lahars, debris avalanches are usually transported dry, with no necessary addition of water.

Other features that had been previously interpreted as mudflows, landslides, or glacial moraines were re-examined in the light of what had happened at Mount St Helens. Already, it seems that any self-respecting strato-volcano must have had its debris avalanche. They have been indicated, for instance, on Etna, Popocatépetl, Colima, Unzen and Galunggung. Shasta in the Cascade Range had an enormous debris avalanche about 300000 years ago that redistributed much of the volcano in an apron around its southern base. It covers 450 km^2, and reaches 26 km^3 in volume. The great eruption of Bezymianny in 1956 was incompletely understood until events at Mount St Helens set them into perspective. At Bezymianny, too, the erup-

61

tion had been unleashed by a blast, followed by a debris avalanche that now covers 98 km² around its base.

Most debris avalanches affect large volcanoes, but they can also take place on domes and cinder cones. In 1650, a debris avalanche occurred on the northern slopes of the Chaos Crags dome on Lassen Peak, forming the extensive and well named zone of dacitic boulders, scraps and ash, called Chaos Jumbles. A debris avalanche has recently been indicated from the cinder cone of Puy de Gravenoire in Auvergne. The puy erupted on the brink of the 200 m scarp bordering the Grande Limagne, 10 km south of Clermont-Ferrand, during the last ice age between 52 000 and 82 000 years ago, when permanently frozen ground covered this area. The development of the avalanche was probably facilitated when rising magma melted the frozen ground. In the early stages of the eruption, a mixture of the granitic basement rocks and both lavas and cinders from the unsupported east flank of the cone swept down the scarp into the Grande Limagne in a swathe 500 m broad. Its total volume of 0.1 km³, commensurate with the small size of the 200 m-high cinder cone, illustrates the enormous differences of scale between this avalanche and that at Bezymianny, for instance.

Lahars or volcanic mudflows

Lahars are amongst the most destructive forces associated with eruptions. Except in very rare cases, such as that expelled from the Soufrière dome in Guadeloupe on 8 July 1976, lahars are not themselves eruptions, but the result of eruptions. They are the scourge of Indonesia – a country of high volcanoes, high rainfall and high population – and it is fitting that the Indonesian word "lahar" has been adopted to describe them. Lahars are a combination of water and unsorted angular fragments of all sizes, from ash and cinders to large boulders, new and old lava, trees, houses and even human remains, sweeping down slope in a roaring mass and leaving a grey trail of devastation in their wake. Their most recent exploit was the destruction of Armero in Colombia. They are generated best on high strato-volcanoes, such as Nevado del Ruiz, where hot erupted fragments either melt ice or snow at the summit, or mix with the waters in crater lakes. A muddy mixture is created that starts moving down slope under gravity, picking up more erupted materials and loose surface debris as it progresses. Lahars travel faster over steep slopes and can thus pick up more material, more easily, especially if they are confined within narrow, steeply inclined valleys, where they can develop the considerable erosive force that enables them to sweep away trees, bridges and homes and exceed speeds of 50 km an hour. In 1980, for instance, the lahars from Mount St Helens severely deformed lorries, and twisted steel bridges across the River Toutle as if they were rubber. However, they rapidly lose their momentum once they encounter gently sloping plains. Had it not been for its lahars, the eruption of Nevado del Ruiz in 1985 would have passed off almost unnoticed, but it was Armero's misfortune to lie just too close to the exit of the River Lagunillas canyon, down which the lahars roared.

The speed of lahars makes them notably lethal. In the present century alone, lahars have killed several thousand times as many people as lava-flows (which seem almost docile in comparison), and they feature high on the list of agents of natural catastrophe during historic time. Nevado del Ruiz has been far from being the only culprit. In 1919, the lahars generated by an eruption in the crater lake of Kelut in Indonesia devastated over $130 km^3$ of agricultural land and killed more than 5000 people in less than an hour. This was the worst of several Indonesian disasters provoked by lahars in the past 400 years: Makian in 1760 killed 2000 people; Papandayan in 1772 killed nearly 3000; Galunggung in 1822 killed 4000; Awu killed 3000 people in 1856 and a further 1500 in 1892.

An eye-witness account of a small lahar survives from the eruption of Montagne Pelée in 1902. At about 12.45 on 5 May the rim of the Etang Sec at the summit gave way, precipitating its waters into the Rivière Blanche. The result was described by Dr Guérin, owner of the sugar and rum factory at the mouth of the stream, who was preparing to leave his home nearby.

> At ten past twelve [sic], people rushed in front of the house shouting out in terror: "The mountain's coming down!" Then I heard a noise that I can't compare with anything else – an immense noise – like the devil on Earth! A black avalanche beneath white smoke, an enormous mass, full of huge blocks, more than 10 m high and at least 150 m wide, was coming down the mountain with a great din. It left the river bed and rolled up against the factory like an army of giant rams. I was rooted to the spot.
>
> My unfortunate son and his wife ran away from it towards the shore and I saw them go out of sight behind the factory. All at once, the mud arrived. It passed 10 m in front of me. I felt its deathly breath. There was a great crashing sound. Everything was crushed, drowned, and submerged. My son, his wife, thirty [sic] people, and huge buildings were all swept away by the waves of the avalanche. Three of these black waves came down one after the other, making a noise like thunder, and made the sea retreat. Under the impact of the third wave, a boat moored in the factory harbour was thrown 150 m over an 8 m high factory wall, killing one of my foremen who was standing next to me. I went down to the shore. The desolation was indescribable. Where a prosperous factory – the work of a lifetime – had stood a moment before, there was now nothing left but an expanse of mud forming a black shroud for my son, his wife and my workmen.

The lahar probably travelled at a speed of 120 km or 160 km an hour and proved so powerful that much of the factory was not so much buried in the mud as literally pushed into the sea. The lower reaches of the Rivière Blanche valley were choked with black mud at least 20 m thick and 300 m wide, which spread 30 m out to sea. The only signs of Dr Guérin's factory were the top of the great scales and the chimney, rising out of 6 m of mud.

But this lahar was far from large; it only devastated a valley 6 km long and amounted to probably no more than 8 million m^3 of material, and it made but a small inaugural contribution to the death-toll of the eruption of Montagne Pelée. The lahar also propagated a little tsunami which swept 20 m over the central shores of St Pierre about 15 minutes later that did nothing to reassure the inhabitants of the doomed city.

Tsunamis

When large eruptions take place in or near the sea, great sea-waves are commonly generated, which are known by their Japanese name of "tsunami". Although they are sometimes caused by explosive eruptions, or by lahars, the more important volcanic tsunamis are propagated most frequently in relation to caldera collapse or debris avalanches. (However, major eathquakes usually form the largest tsunamis of all.) A tsunami usually spreads out in a broad fan-shaped sector, directed by the event that creates it. Tsunamis neither change direction easily, nor turn round headlands, and thus not all areas nearby are affected equally. In deep open waters, tsunamis are scarcely noticeable, but the waves increase in both size and speed as the water becomes shallower. They can crash onto the shore at speeds of 250 km an hour in walls of water over 30 m high, devastating settlements and drowning many of their inhabitants. Tsunamis have been responsible for about a quarter of all the deaths caused by eruptions in historic time.

Many have thought that the tsunami from Santorini was a major contributor to the sudden collapse of the Minoan civilization of Crete, although this now seems unlikely. A tsunami from the same eruption has also been invoked to explain the drowning of the Egyptian armies pursuing Moses and the Israelites across the "Red Sea" – most probably, in fact, the "Sea of Reeds" in the northeastern part of the Nile delta. In Japan, 14524 people were killed when a debris avalanche from Unzen crashed into the sea on 21 May 1792 and propagated a tsunami, between 35 and 55 m high that flooded the adjacent coast. At Krakatau, on 27 August 1883, the collapse of the caldera (or, perhaps, a debris avalanche, or a nuée ardente, or all three) generated tsunamis reaching a height of 30 m that inundated the adjacent coastal areas of Java and Sumatra and drowned about 36000 people. Many coastal settlements on both islands were almost obliterated, and a ship was transported 3 km inland. Excluding the effects of the famine after the eruption of Tambora in 1815, the tsunamis from Krakatau still represent the record for deaths registered through volcanic activity during historic time.

Gas eruptions

Gas is an almost inevitable component of eruptions and the major element in their explosions. On occasions, gas emissions take place without explosions and escape from fissures and vents and help to form fumaroles, solfataras and mudpots. Very rarely, such emanations are large enough to cause loss of life: on 20 February 1979, the emission of $0.1 km^3$ of carbon dioxide gassed 142 people on the Deng Plateau in Java. Larger emanations, with relatively small explosions, have also taken place. The most notable in historic time occurred during the Lakagígar eruption of 1783 that caused the great famine in Iceland. Latterly, emissions of carbon dioxide from old volcanic lakes have sprung to the forefront as a result of recent events in Cameroon.

On 15 August 1984, carbon dioxide was expelled from Lake Monoun in western Cameroon, but, although 37 people were killed, the eruption created little scientific stir. However, on 21 August 1986, carbon dioxide was also emitted from Lake Nyos which occupied a similar maar nearby (Fig. 5.13). Carbon dioxide is all the more dangerous because it is invisible, and, in pure masses, hugs the ground instead of dissipating in the air, killing any human or animal unfortunate to inhale it within a few minutes.

Views differ considerably about the origin of this gas, with the protagonists divided on national lines. It may have slowly exuded from old volcanic fissures and accumulated at the bottom of the lake over a long period, dissolved or trapped at depth by the cold, dense layers of water. Then, possibly as a result of a landslide or a small earthquake, or even because of recent heavy rainfall in the wet season, the gas suddenly escaped to the surface and spilled beyond the lake. However, others believe that the phenomena indicated a volcanic gas eruption, which seems to be supported by the eye-witness accounts. Preliminary emissions of gas started from Lake Nyos at 16.00 on 21 August 1986, followed by an earth-tremor at 20.00, which was succeeded by intermittent hissing gas jets and explosions like gunfire from the lake. A major explosion was heard at 22.00 on 21 August; and the carbon dioxide escaped northwards from the lake and progressed down valley on its lethal errand at a speed of nearly 50 km an hour. Between $0.6 km^3$ and $1 km^3$ of carbon dioxide was expelled, and high concentrations of the gas were maintained for about 23 km. Survivors on the edge of the invisible cloud smelled rotten eggs or gunpowder; some felt burning acidic raindrops on their skin; others felt the ground cracking; many who inhaled only a few breaths of gas became confused or lost consciousness; some even walked into the gas and died before they could escape from it. None of the people or animals enveloped in the gas for more than a few minutes survived. About 6000 cattle died. In Nyos village, 600 out of the 606 inhabitants were killed. In all, the carbon dioxide asphyxiated perhaps as many as 1742 people.

Hydrothermal eruptions

Hydrothermal activity causes the least explosive of all volcanic eruptions and forms hot springs, geysers, fumaroles, solfataras and mudpots. Hot springs give off little gas, but may precipitate a range of mineral compounds at the surface; geysers erupt columns of water and steam at regular intervals; mudpots contain more precipitates, some gases and much less water than geysers; and fumaroles and solfataras give off gases and steam. Lava is never emitted and the thickness of the products rarely exceeds 20 m. These are small-scale, often beautiful eruptions, affecting little but the

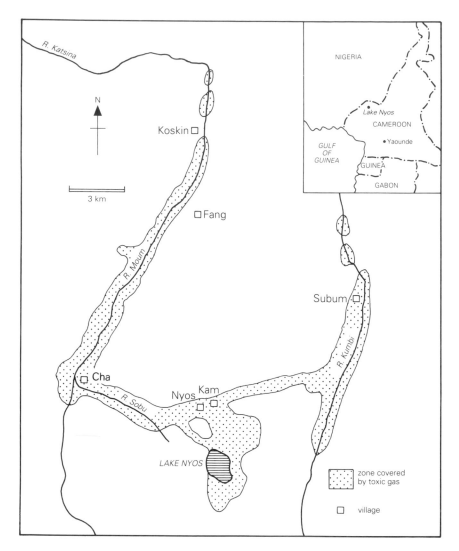

Figure 5.13 Emission of carbon dioxide from Lake Nyos, Cameroon, 21 August 1986.

area where they occur. They are subject to frequent changes because they are easily disturbed by earth-tremors, variations in atmospheric precipitation, or when visitors conduct experiments and bombard them with coins or beer-cans as if they were the Trevi Fountain.

Hydrothermal activity occurs most commonly where abundant precipitation or snowmelt infiltrates deeply fissured ground that is heated by shallow magma. As a result, temperatures can rise by as much as 1°C for every 10m of depth below the surface. The conduits facilitating downward water seepage and lateral flow also help its return, heated, under hydrostatic pressure. The circulation system can be relatively open, when hot water can bubble to the surface more or less continuously, as in Boiling Lake in Dominica in the West Indies, and in Furnas caldera in São Miguel in the Azores (Fig. 5.14), or it can be restricted so that hot water reaches the surface only intermittently, often in the form of geysers. If the water circulates only slowly and temperatures are high, chemical reactions often take place with the enclosing rocks, whose altered fragments are then mixed with, or dissolved in, the water. Although most volcanic areas display some hydrothermal activity, it is best developed at present in Iceland, the North Island of New Zealand, and in Yellowstone Park.

Geysers are probably the most beautiful example of hydrothermal interplay. If the circulation system is restricted, and the ground water is retained for a while within subterranean cavities, then the emissions can burst out only intermittently. Infiltrated water can be heated by magma underneath to over 200°C without boiling. It tends to rise, however, so that the decrease in pressure allows steam to form which pushes the water upwards. Water and steam then gush up the vent and burst forth

Figure 5.14 Hot spring in Furnas caldera, São Miguel, Azores.

Figure 5.15 Beehive Geyser, Yellowstone Park, Wyoming,
in one of its infrequent eruptions.

at intervals onto the surface as geysers. The gushing stops when the water supply is
exhausted, but the process is repeated as rain and snowmelt replenish the conduits.
In Yellowstone Park these conditions are admirably fulfilled; the temperature is
200°C at a depth of 200m and the magma itself lies only about 6km below the surface. Each geyser develops a regime of its own dictated by its water supply and its
fissure and cavity pattern. Some geysers, such as Old Faithful, are so reliable that a
timetable of their displays is provided for visitors. It rewards them by producing columns of steam and water rising 20m or more at intervals, at present, of 67 minutes.
Clepsydra is another geyser which, as its name suggests, erupts regularly like a water
clock. Other geyser activity is more impressive, although more episodic in character, and some, such as the aptly named Grotto and Beehive geysers, also deposit
whitish silica on the surface in hooded masses around their conduits (Fig. 5.15).

Iceland, which gave the word "geyser" to the international vocabulary, now has
only about a dozen active geysers, which, in fact, would only operate sporadically
without artificial assistance. The area around Geysir is amongst the main thermal
centres in the country, with springs ranging between 80°C and 100°C. The Great
Geyser itself is now waning after about 8000 years of activity. It erupted every half
hour 200 years ago, but nowadays it only bursts out after long, irregular intervals
unless it is stimulated by soap-powder. Strokkur, nearby, is kept on a bore-hole life-
support machine and can thus still spurt 20m into the air every few minutes so as not
to disappoint the customers.

Hot springs have a quicker, less restricted water circulation and can rise to the surface without gas or steam. Sometimes the hot waters pass through limestones as well as volcanic rocks and bring up calcium carbonate, which is then redeposited. They form the decorative organ-pipe terraces at Pamukkale in central Turkey and at Mammoth Springs in Yellowstone Park (Fig. 5.16). In Yellowstone Park these springs also form superb pools, where the varying temperatures in the still water support different sorts of algae, each with a distinctive brilliant colour. Grand Prismatic Pool and Beauty Pool contain rings of red, yellow, emerald green or sapphire blue, and even the streamlets draining from them are multicoloured.

Hot springs become notably muddy when the water table is lower, where precipitation is rare, or where the waters are contaminated by precipitates held in solution or suspension (Fig. 5.17). They form mudpots or mud volcanoes that are usually grey, although they can be brilliantly coloured like Fountain Paint Pot in Yellowstone Park. They draw their fascination from their constant bubbling that can be as impressive as the beginning of Macbeth. Yellowstone Park has spectacular mudpots, ranging from 1 m to 20 m across, that have been given picturesque names such as Black Dragon's Caldron and Cooking Hillside.

Fumaroles are formed when only steam and hot gases are expelled from small vents commonly less than 10 cm across. They are often very numerous, and Yellowstone Park alone has thousands. When assembled in large fuming hollows, they often look more threatening than they really are, and have inspired such exaggerated place names as Colter's Hell at Cody, Wyoming, Devil's Kitchen, or Bumpass Hell on Lassen Peak. Nevertheless, fumaroles are capable of burning and scalding. Those issuing onto the Acqua Calda beach at Vulcano, for instance, are hot enough to scald the sensitive parts of unwary bathers. Most active volcanoes in repose frequently show fumarole activity. Whatever the location of the fumaroles, the water comes mainly from recycled atmospheric precipitation and only rarely from the rising magma. Some fumaroles are as hot as 1000°C and many exceed 100°C. Their gas content often varies according to the temperature. Relatively cool fumaroles between 100°C and 300°C often give off hydrogen sulphide that reacts with the oxygen in the air to produce sulphur deposits. These are known by the Italian name of solfataras.

La Solfatara itself lies in the Phlegraean Fields west of Naples. Its crescent-shaped cone is wrapped around the Piano Sterile, the flat-floored crater, some 750 m across, containing about 25 active little vents, from which escape hot gases such as carbon dioxide, sulphur dioxide and hydrogen sulphide, as well as mud, steam and water. This activity has probably continued ever since the crater was formed about 4000 years ago. The most important episode in historic time occurred in 1198 when a big mud geyser spouted 10 m into the air for several days after a small earthquake. A similar event took place on 23 July 1930. At present, the Piano Sterile is a wide arena, covered with grey mud, into which a number of bubbling muddy pools are sunk. Many of the solfataras are encrusted with beautiful, if fragile, sulphur crystals. This sulphur, however, reacts with water to produce sulphuric acid, which alters the rocks of the crater floor and forms the white "bianchetto" coating the drier surfaces. The activity of La Solfatara is dominated by the recycling of rainwater, heated to

Figure 5.16   Mammoth Springs Terraces, Yellowstone Park, Wyoming.

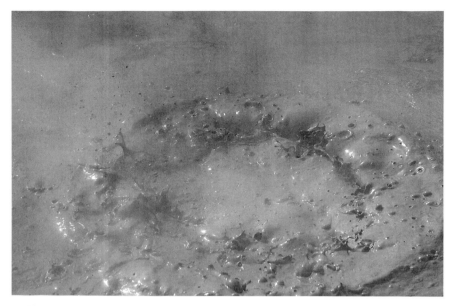

Figure 5.17 A bubbling mud-pot, near Reykjavík, Iceland (courtesy Margot Watt).

high temperatures, which reach 120°C at a depth of 150m. The predominance of gas and mud is no doubt related to the high subterranean temperatures and the relative scarcity of atmospheric water. Each cluster of vents has its own particular characteristics. The Forum Vulcani is a vigourous steam vent; the Bocca Grande has the hottest emissions at a temperature of 185°C; in the centre of the crater, Fangaia ("mire") has hot mudpots usually about 2m deep. In the active central parts of the crater, hardened mud forms grey mounds between the pots and pools, but the crust is sometimes dangerously thin, resonates underfoot, and masks boiling waters just below the surface. In contrast, the northwestern area of the Piano Sterile is solid enough to support a campsite and bar for those who appreciate sulphurous odours with their leisure.

In several countries, hydrothermal activity has been exploited for heating. Chaudes Aigues ("hot waters") in Auvergne, runs a central-heating system from its hot springs; and those at Larderello in Central Italy, and near Ribeira Grande in São Miguel in the Azores, are used for industrial and domestic power. In Iceland, boreholes reaching 2000m in depth have been sunk to tap the hot water to provide cheap, clean, ecological heat for homes, swimming pools, industrial and commercial installations, and greenhouses. All Reykjavík is heated in this way. Electricity is generated from the hydrothermal heat of Krafla volcano. At Hveragerdi, the geothermally heated greenhouses, covering 10ha, have become world-famous because bananas, as well as less exotic tomatoes, have been ripened almost within reach of the Arctic Circle. Iceland, inside and underground, can be quite warm . . .

These hydrothermal emissions are often symptoms of waning volcanic activity,

71

or simply mark periods of repose between more violent explosions. However, many volcanic areas are so riddled with hydrothermal fissures and conduits that the whole rock is altered and weakened. Much of El Piton, the summit cone of Teide, for instance, has been transformed into a powdery white mass resembling the "bian-chetto" at La Solfatara. But, by way of compensation, hydrothermal alteration of rhyolite in Wyoming formed the "yellow stone" that gave its name to a beautiful volcanic region. Not least of its attractions is the superb canyon cut through this golden stone by the Yellowstone River.

6

Eruptions

Scarcely a day goes by without a volcanic eruption occurring somewhere in the world. Every century in historic time, and perhaps every century since the beginning of the Earth, has had its share of major volcanic eruptions. Many will have happened in remote, inaccessible places and have never been recorded. Others have affected populated areas and forced themselves into the legends or the archives of mankind because of the damage and deaths that they have caused. The present century has been no exception. It was inaugurated with tremendous outbursts at the Soufrière in St Vincent, at Montagne Pelée in Martinique, and Santa Maria in Guatemala in 1902. Mid-century was celebrated a little late with the great eruption of Bezymianny in Kamchatka in 1956, and, since Mount St Helens erupted in 1980, three hitherto little-known volcanoes, El Chichón, Nevado del Ruiz and Pinatubo, have achieved world renown. Statistically, at least one other large eruption should take place before the century ends.

The eruptions not only vary in their style and power, but also in their effects and interaction with the natural and the human environment. Each eruption thus has a different scientific and social significance, which gives a major added dimension to the study of volcanoes. Several eruptions have been selected to illustrate these different interactions with the environment. The eruption of Vesuvius in AD 79 was probably the most violent in Europe during historic time; it also became one of the most famous because it destroyed and preserved the Roman city of Pompeii and was also described by an intelligent eye-witness. Its archaeological, as well as its volcanological, significance can scarcely be exaggerated. In recent years, three major congresses investigating the Bronze Age eruption of Santorini, and its possible implication in the collapse of the Minoan civilization in nearby Crete, have also linked volcanology and archaeology. The fissure eruption of Lanzarote between 1730 and 1736 can be re-interpreted in the light of newly discovered archives as well as in relation to the eye-witness account of a local priest. The great killer eruptions of the present century have exercised a macabre, and perhaps more immediate, fascination. Montagne Pelée annihilated St Pierre in Martinique in five minutes in 1902, and revealed to science for the first time the devastating rôle of nuées ardentes. The Nevado del Ruiz eruption in Colombia in 1985 demonstrated that a volcano can also be lethal from afar when it forms lahars. Mount St Helens, surely the most closely studied volcano of the 1980s, illustrated the effects of blasts and debris avalanches whose power had hardly been previously fully comprehended. The much smaller eruption of Heimaey off Iceland in 1973 represents a relatively rare case of a sort of victory of humanity over a volcanic onslaught – based, it must be said, on a certain amount of luck in the location of the eruptive vent, as well as upon excellent disaster planning. The formation of Capelinhos, off Faial in the Azores in 1957–8, illustrated the varied and rapidly changing relationships with the natural environment, ranging from the influx of sea water into the vent to subsequent partial destruction of the volcano by marine erosion. The eruption of Pinatubo in 1991 is the most recent large-scale outburst that took place from a volcano that was believed to be probably extinct. These accounts are only a personal selection: the choice available is large.

Vesuvius, AD 79

Vesuvius is the focus of one of the world's most famous and beautiful landscapes. The view from the Via Partenope in Naples has been painted and photographed so often that Vesuvius has been instantly recognizable for centuries. Its serenity in repose, however, belies the vehemence of its temper. Vesuvius is the most violent and dangerous volcano erupting regularly in Europe and it has become an archetype of lethal volcanic activity. Its outbursts, recorded with varying degrees of fantasy and accuracy for more than 2000 years, have repeatedly destroyed the crops, property and sometimes the lives of those enticed to its flanks by the renowned fertility of its soils.

The eruption in the reign of the Emperor Titus that buried Pompeii and Herculaneum made them the most famous of archaeological sites and also made Vesuvius the most celebrated volcano on Earth (Fig. 6.1). It is the subject, too, of the first account of scientific value of any volcanic eruption, which was given by Pliny the Younger in two letters to the historian Tacitus. The eruption in AD 79 was the grandest of all the historical outbursts of Vesuvius and it has become a prototype of powerful eruptions called Plinian in honour of Pliny the Younger and his uncle, Pliny the Elder, who was the most distinguished victim of the catastrophe. All these archaeological, historical, geological and cultural associations have given this particular volcanic disaster a glamour that perhaps no other can match.

In AD 79 Pompeii was one of the largest towns in Campania, with a population of about 20 000 (Fig. 6.2); it lay 10 km southeast of Vesuvius. Herculaneum was a smaller seaside town situated 7 km due west of the mountain. Although classical scholars such as Strabo had recognized its volcanic nature, Vesuvius had been calm for centuries and there was no apparent reason to suppose that it presented any threat whatsoever to one of the most idyllic places in the Roman world. The main perceived hazard in the area was earthquakes. Most caused no more than a shudder in the ground and people alike, but a larger earthquake on 5 February AD 62 had so severely damaged Pompeii that the city still had not been fully repaired 17 years later. The eruption in AD 79 was thus totally unexpected and was also of an intensity

Figure 6.1　Vesuvius from the Forum at Pompeii.

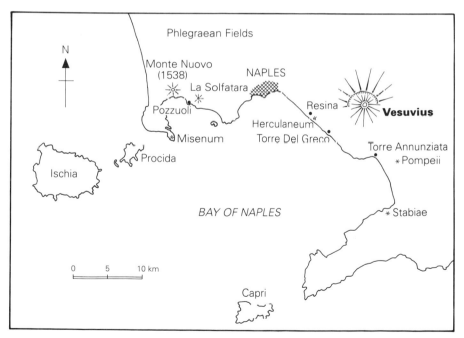

Figure 6.2　Vesuvius and the Bay of Naples.

that has not since been equalled in Europe. Pliny the Elder was a scientist and author of a famous natural history, as well as being a commander of the Roman fleet. His nephew, not yet 18 at the time of the eruption, himself later became a noted writer. When the eruption began on 24 August AD 79, Pliny the Elder, his sister and her son, Pliny the Younger, were staying at Misenum, the naval base 32 km from Vesuvius at the western end of the Bay of Naples. Pliny the Elder died at Stabiae, across the Bay, 17 km south of Vesuvius, on the morning of 25 August, most probably either from a heart attack brought on by his exertions in the suffocating atmosphere, or by inhaling volcanic ash and toxic gases. Meanwhile, Pliny the Younger and his mother wandered amidst a terrified crowd in the stifling darkness that persisted throughout 25 August in and around Misenum until the eruption ended early on 26 August. None of them ever went to Pompeii during the eruption and Pliny the Younger does not mention the city in his narrative. But his accounts – both slightly abridged here – of his uncle's death in the first letter, and his own nightmare experiences at Misenum in the second, have retained their vivid impact and reveal some of the horror of a really powerful eruption. Listen.

My uncle was at Misenum in command of the fleet. On 24 August, about one in the afternoon, my mother pointed out a cloud with an odd size and appearance that had just formed. From that distance it was not clear from which mountain the cloud was rising – although it was found afterwards to be Vesuvius. The cloud could best be described as more like an umbrella pine than any other tree, because it rose high up in a kind of trunk and then divided into branches. I imagine that this was because it was thrust up by the initial blast until its power weakened and it was left unsupported and spread out sideways under its own weight. Sometimes it looked light-coloured, sometimes it looked mottled and dirty with the earth and ash it had carried up. Like a true scholar, my uncle saw at once that it deserved closer study and ordered a boat to be prepared. He said that I could go with him, but I chose to continue my studies.

Just as he was leaving the house, he was handed a message from Rectina, the wife of Tascus, whose home was at the foot of the mountain and who had no way of escape except by boat. She was terrified by the threatening danger and begged him to rescue her. He changed plan at once and what he had started in a spirit of scientific curiosity, he ended as a hero. He ordered the large galleys to be launched and set sail. He steered bravely straight for the danger zone which everyone else was leaving in fear and haste, but still kept on noting his observations.

The ash already falling became hotter and thicker as the ships approached the coast and it was soon superseded by pumice and blackened burned stones shattered by the fire. Suddenly the sea shallowed where the shore was obstructed and choked by debris from the mountain. He wondered whether to turn back as the Captain advised, but decided instead to go on. "Fortune favours the brave", he said, "take me to Pomponianus". Pomponianus lived at Stabiae across the Bay of Naples, which was not yet in danger, but would be threatened if it spread. Pomponianus had already put his belongings into a boat to escape as soon as the contrary, onshore wind changed. This wind, of course, was fully in my uncle's favour and quickly brought his boat to Stabiae. My uncle calmed and encouraged his terrified friend and was cheerful, or at least pretended to be, which was just as brave.

Meanwhile, tall, broad flames blazed from several places on Vesuvius and glared out through the darkness of the night. My uncle soothed the fears of his companions by saying that they were nothing more than fires left by the terrified peasants, or empty abandoned houses that were blazing. He went to bed and apparently fell asleep, for his loud, heavy breathing was heard by those passing his door. But, eventually, the courtyard outside began to fill with so much ash and pumice that, if he had stayed in his room, he would never have been able to get out. He was awakened and joined Pomponianus and his servants who had sat up all night. They wondered whether to stay indoors or go out into the open, because the buildings were now swaying back and forth and shaking with more violent tremors. Outside, there was the danger from the falling pumice, even if it was only light and porous. After weighing up the risks, they chose the open country and tied pillows over their heads with cloths for protection.

It was daylight everywhere else by this time, but they were still enveloped in a darkness that was blacker and denser than any night and they were forced to light their torches and lamps. My uncle went down to the shore to see if there was any chance of escape by sea, but the waves were still running far too high. He lay down to rest on a sheet and called for drinks of cold water. Then, suddenly, flames and a strong smell of sulphur giving warning of yet more flames to come, forced the others to flee. He himself stood up, with the support of two slaves and then suddenly collapsed and died, because, I imagine, he was suffocated when the dense fumes choked him. When daylight returned on the following day, his body was found intact and uninjured, still fully clothed and looking more like a man asleep than dead.

★ ★ ★

Meanwhile, my mother and I had stayed at Misenum. After my uncle left us, I studied, dined and went to bed, but slept only fitfully. We had had earth-tremors for several days, which were not especially alarming because they happen so often in Campania. But that night they were so violent that everything felt as if it were being shaken and turned over. My mother came hurrying to my room and we sat together in the forecourt facing the sea.

By six o'clock the dawn light was still only dim. The buildings around were already tottering and we would have been in danger in our confined space if our house had collapsed. This made us decide to leave town. We were followed by a panic-stricken crowd that chose to follow someone else's judgement rather than decide anything for themselves. We stopped once we were out of town and then some extraordinary and alarming things happened. The carriages we had ordered began to lurch to and fro although the ground was flat, and we could not keep them still even when we wedged their wheels with stones. Then we saw the sea sucked back, apparently by an earthquake, and many sea-creatures were left stranded on the dry sand. From the other direction over the land, a dreadful black cloud was torn by gushing flames and great tongues of fire like much-magnified lightning.

The cloud sank down soon afterwards and covered the sea, hiding Capri and Cape Misenum from sight. My mother begged me to leave her and escape as best I could, but I took her hand and made her hurry along with me. Ash was already falling by now, but not very thickly. Then I turned round and saw a thick black cloud advancing over the land behind us like a flood. "Let us leave the road while we can still see", I said, "or we will be knocked down and trampled by the crowd". We had hardly sat down to rest when the darkness spread over us. But it was not the darkness of a moonless, or cloudy night, but just as if the lamps had been put out in a completely closed room.

We could hear women shrieking, children crying, and men shouting. Some were calling for their parents, their children, or their wives, and trying to recognize them by their voices. Some people prayed for death because they were so frightened of dying. Many begged for the help of the gods, but even more imagined that there were no gods left and that the last eternal night had fallen on the world. There were also those who added to our real perils by inventing fictitious dangers. Some claimed that part of Misenum had collapsed, or that another part was on fire. It was untrue but they could always find somebody to believe them.

A glimmer of light returned, but we took this to be a warning of approaching fire rather than daylight. But the fires stayed some distance away. The darkness came back and ash began to fall again, this time in heavier showers. We had to get up from time to time to shake it off or we would have been crushed and buried under its weight. I could boast that I never expressed any fear at this time, but I was only kept going by the consolation that the whole world was perishing with me.

After a while, the darkness paled into smoke or cloud, and the real daylight returned but the sun was still shining as wanly as during an eclipse. We were amazed by what we saw because everything had changed and was buried deep in ash like snow. We went back to Misenum and spent an anxious night switching between hope and fear. Fear was uppermost because the earth-tremors were still continuing and the hysterics still kept on making their alarming forecasts. But, by then, my mother and I had no intention of leaving the house until we got news of my uncle.

The changes wrought by the eruption were even greater than Pliny described. It was Pompeii, however, 10km southeast of Vesuvius, that bore the brunt of the eruption, and it suffered much more than Misenum because the northwesterly wind blew ash and pumice over it for 18 hours. Unknown to the Pompeiians, a nuée ardente had also destroyed Herculaneum the previous evening, and other nuées were to devastate neighbouring towns before the eruption ended. Pompeii and Herculaneum then remained entombed until they achieved immortality after excavations were initiated in 1748. The early excavations were scarcely more than treasure hunts. In 1820, for instance, the French poet Lamartine was offered a pick, which he promptly gave to three young girls, to excavate a Pompeiian house. Later, Fiorelli developed the technique of making plaster casts of the remains of the victims before the enveloping ash was disturbed, thus preserving their outlines for posterity. As scientific techniques of excavation advanced, the sheer variety of the relics brought to light has made Pompeii and Herculaneum perhaps the greatest single sources of everyday Roman culture. In recent years, too, the accuracy of Pliny the Younger's account has been fully appreciated and it can now be closely linked with the archaeological and geological evidence revealed on the sites.

From 20 August, earth-tremors of increasing frequency centred on Vesuvius gave premonitory, but unheeded signs of the impending eruption. On the morning of 24 August, the initial phreatic explosion cleared the conduit and was sufficiently disturbing to induce Rectina to send to Pliny the Elder for assistance. The next phase had just begun when her messenger reached Misenum.

It was 13.00 on 24 August when the Plinian eruption began. Its column was seen from Misenum rising into the stratosphere, with its upper parts branching outwards like a pine (*Pinus pinea*). The upward self-sustaining thrust of gas and fragments soon reached a height of 27 km, while the upper winds winnowed out the ash and pumice and rained them down southeastwards beyond Stabiae. Pompeii was situated in the zone of maximum accumulation, and throughout the next seven hours the city was covered by about 1.40 m of white ash and pumice that caused roofs to collapse under its weight. Probably by the evening most of the Pompeiians had panicked, fled, and saved their lives. Meanwhile, pumice was falling so heavily in the afternoon that Pliny the Elder could not land near Rectina's house and was forced to sail to Stabiae and death.

From about 20.00 on 24 August, grey, coarser fragments were erupted in a column that rose to a height of 33 km as the eruption reached its climax and covered a wide area southeast of Vesuvius. However, the western areas around Misenum remained unaffected on 24 August and Herculaneum, much closer to Vesuvius, had been spared all but showers of fine ash. But not for long.

At 01.00 on 25 August came the major change in the eruption, when a series of lethal nuées ardentes began to form as the lower column collapsed from time to time, crumbling like a pillar of fire.

The first nuée ardente, at 01.00, was funnelled westwards at a speed of at least 100 km an hour and reached Herculaneum in less than 4 minutes. Its remaining inhabitants had time only to flee to the shore, where they died choking in the hot swirling toxic cloud that engulfed the town. Hundreds of victims were discovered on the coast in excavations in 1982 and many more may be unearthed as they proceed. As less than a dozen victims had previously been discovered, it was assumed that the people of Herculaneum had been able to flee to safety, and thus that the town must have been among the last casualties of the eruption. Herculaneum, in fact, seems to have been the first major victim of Vesuvius in AD 79.

Meanwhile, as the grey pumice continued to fall over Pompeii, a second nuée ardente at about 02.00 completed the burial of Herculaneum. A third nuée ardente at about 06.30 spread as far as the northern walls of Pompeii, and may have mobilized the many hundreds of inhabitants still remaining in the city. A great increase in earth-tremors about that time can only have reinforced their fears. By now, Pompeii was already blanketed by 2.4m of ash, pumice and even larger rock fragments that were still raining down in the stifling pitch darkness. Those that tried to flee stirred up clouds of hot dust as they waded through the fragments choking the streets. The same conditions reigned at Stabiae, where Pliny and his friends were afraid of being trapped in their rooms by the rapidly thickening blanket of pumice as the night progressed.

At 07.30 on 25 August, the vertical impulse of the eruption waned once more, and the resulting collapse of the column created a fourth nuée ardente which completely overwhelmed Pompeii and flattened most of its upper storeys. All the 2000 inhabitants remaining in the city suffered a terrible fate. Some were killed by falling columns or tiles and bricks ripped from the buildings, others were poisoned by the toxic gases, yet others were baked, but most were asphyxiated by a mixture of hot ash and mucus that they inhaled with their last breaths. Many suffocated clutching their most treasured possessions, writhing in breathless agony, trying to keep the hot ash, but not the stifling air, from their mouths. A dog died on its lead, gladiators in their irons, a doctor with his medicine bag, and an athlete with his bottle of oil. Hundreds perished in the useless shelter of their kitchens, where they were entombed with eggs they had stored and cakes they had just baked. Two small boys holding hands died together as they tried in vain to protect themselves beneath a tile.

The *Dies Irae* was over for the Pompeiians at 07.30, but not for Vesuvius. A fifth and even larger nuée ardente swept again over Pompeii five minutes later and carried on to the outskirts of Stabiae. Earth-tremors reached a frightening pitch throughout the region. About 08.00 on 25 August the sixth and largest nuée ardente swept down as far as Stabiae, where it killed Pliny the Elder as his friends fled southwards with cushions tied over their heads. Another branch of the same nuée ardente surged westwards across the waters of the Bay of Naples and halted within sight of Pliny the Younger and his mother in the panic-stricken crowd on the hills above Misenum. Ash rained down upon them, the ground shook and it seemed as if the last eternal night had fallen upon the world. Some were so afraid of dying that they prayed for death – and they were 32km from Vesuvius!

Vesuvius released other nuées ardentes throughout the pitch-black morning of 25 August. Then, as the day went on, the nuées ardentes diminished in vigour, the erupting column was reduced in height, and the eruption waned. Eventually, glimmers of daylight returned to the more distant areas such as Misenum in the west, the earth-tremors declined; and Pliny the Younger and his mother returned home. On the morning of 26 August when the great eruption drew to a close and daylight at last returned, the whole of Campania was covered with a thick blanket of fine, grey pumice in undulating dunes. About 300 km² around Vesuvius was completely devastated and most of the farms, villages, towns and cities in the vicinity had vanished. Vesuvius had lost 700 m from its summit and the ancient cone had been replaced by a fuming chasm, fringed by the jagged ridge now known as Monte Somma. It was in this chasm several centuries later that Vesuvius began to build its present cone. No lava-flows had been emitted, but, in less than two days, Vesuvius had discharged 1 km³ of white pumice, 2.6 km³ of grey pumice, 0.37 km³ of nuée ardente deposits and 0.16 km³ of other rocks. The eruption had thus delivered at least 4.13 km³ of materials, together with Neptune knows how much debris that also fell into the Bay of Naples.

El Chichón, 28 March 1982

El Chichón is a lonely figure amongst Mexican volcanoes, lost in the tropical forests of the state of Chiapas in southeastern Mexico (Fig. 6.3). Scarcely known before it erupted in 1982, El Chichón became famous virtually overnight, for it was not only comparable in violence to the major Mexican eruptions of historic time such as Jorullo in 1759, San Martin in 1793 and Colima in 1818, but its effects on the global climate were perhaps the greatest since the eruption of Krakatau in 1883. Before 1982, it rose 1350 m above sea level with an oval crater 1900 m by 900 m in size, occupied by a lava dome 400 m high and partly filled by a lake. The summit of the volcano gave off fumaroles and hot springs to show that it was not entirely extinct, but the thick vegetation cover on its flanks indicated that it had not erupted for some centuries. El Chichón had two great eruptions during the Mayan period about AD 700 and about AD 1300, but neither was chronicled in the surviving Maya texts. A series of local earthquakes had also occurred in 1930, but no eruption ensued.

The eruption of 1982 more than compensated for the delay. The first premonitory symptoms were felt during the autumn of 1981, when the local villagers noticed increasing earth-tremors, fumarole activity and a pervading smell of rotten eggs, indicating emissions of hydrogen sulphide. This activity increased during the winter of 1982, but the Provincial Governor and some geophysicists consulted at the time apparently reassured the local population. During March 1982, however, the pace and intensity of the tremors increased and their focus rose from a depth of 5 km to only 2 km below the volcano. These tremors stopped on 28 March at 21.30.

The calm lasted two hours. At 23.30 on 28 March 1982 a great explosion pulverized the dome blocking the vent and produced a column of gas and fragments which rose 20 km into the air and eventually spread out over a distance of 100 km (Fig. 6.3). The weight of the ash that rained down on the villages around El Chichón caused many rooftops to collapse and thousands of people fled. The eruption lasted for six hours. A period of calm ensued for a week, punctuated only by minor explosions, which induced many villagers to return to their wrecked homes.

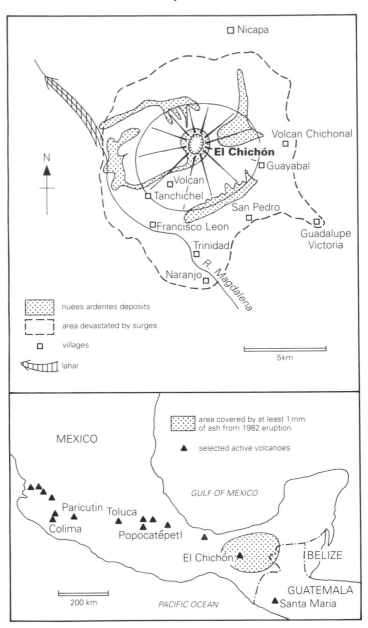

Figure 6.3 Map of El Chichón and the chief volcanoes of Mexico.

On the evening of 3 April, at about 19.30, El Chichón began an even greater eruption, lasting more than four hours. First, a nuée ardente surged in all directions down from the summit and spread 8 km from the crater like the billows extending from the base of a nuclear explosion. It was followed by a column of gas and ash that rose 24 km into the stratosphere. Part of the column collapsed and formed nuées ardentes that swooped down the flanks of the volcano at speeds of up to 350 km an hour in a typical mixture of gas, ash, pumice and blocks of all sizes, at temperatures ranging from 300°C to 600°C. Over 100 km^2 was devastated. The final eruption began at 05.22 on 4 April and repeated the features displayed the previous evening, but with a rather reduced intensity. This ultimate outburst lasted until noon, although darkness reigned in the area for several days afterwards. Minor eruptions continued until 9 April.

When the eruption ceased, a new crater, 1 km across and 230 m deep, had been formed in a summit which was now only 1060 m above sea level. El Chichón had erupted just over 1 km^3 of fragments. Ash had blanketed 50 000 km^2, especially to the east of the vent, and nine villages within a 7 km radius of the volcano were destroyed. In some villages, only fragments of the church, a basketball post, or charred cooking utensils remained as evidence of their former existence. At least 2000 lives and most of the local livestock were lost. Cocoa, coffee and banana plantations were devastated, and nearer the volcano all the vegetation had vanished. More than 20 000 people were made homeless. Meanwhile a bubbling lake of hot water and sulphuric acid had filled a crater that was bordered by fumaroles reaching temperatures above 440°C.

The eruptions of El Chichón formed a large cloud composed of millions of tonnes of fine ash and dust and an unprecedented amount of sulphuric acid. The cloud travelled westwards in the stratosphere, and its first trip around the world took only three weeks. It persisted for many months and by the end of the year had enveloped the whole globe. Its effects lasted for three years. The particles reflected solar radiation and prevented some of the Sun's heat from reaching the Earth. As a result, mean annual world temperatures may have been reduced by 1°C. El Chichón was held responsible for the abnormally cold winters in the United States in 1983–4 and in Europe in 1984–5.

Nevado del Ruiz, 13 November 1985

The very height and remoteness of the glistening ice-capped Andean volcanoes makes them seem almost unreal. But they are real enough – and dangerous: in a single day, Nevado del Ruiz caused the greatest volcanic catastrophe since Montagne Pelée destroyed St Pierre in 1902. The eruption on 13 November 1985 was neither large nor violent for a subduction zone volcano, but it caused damage estimated at US$1000 million. The materials erupted killed no-one directly. Most of the 25 000 deaths were caused by lahars some 60 km from the crater (Fig. 6.4). It is said that almost all the fatalities could probably have been avoided.

Nevado del Ruiz was built to a height of 5389 m by several eruptive episodes during and after the last ice age. The chief active vent, the Arenas crater, occupies a crucial position on the northeastern fringe of an ice-cap 21 km^2 in area, making the steep valleys radiating northeastwards from the summit especially vulnerable during

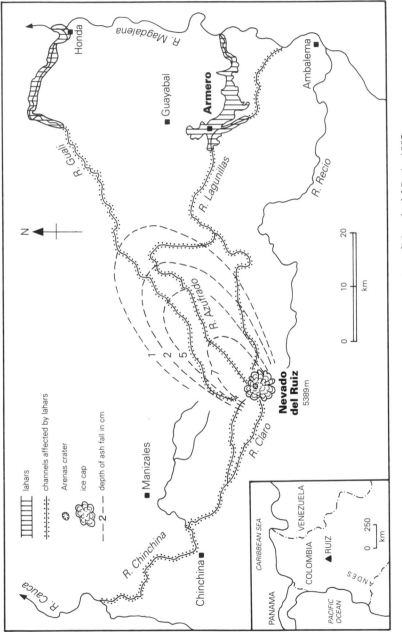

Figure 6.4 Lahars and ashfalls from the eruption of Nevado del Ruiz, 1985.

an eruption. Indeed, in 1985, all the ingredients of a catastrophe were already assembled. Although any lava-flows emitted would be too small to threaten the distant settlements, nuées ardentes and lahars could be particularly menacing. It only required an eruption of hot fragments or nuées ardentes to melt enough waters from the ice-cap to generate unstoppable lahars from the abundant loose materials strewn on the surface by previous eruptions. The valleys of the River Guali, River Azufrado and River Lagunillas would then guide them to the populous plains bordering the River Magdalena, nearly 5000m below and 60km away.

There were two clear precedents. On 12 March 1595, Nevado del Ruiz threw out blocks and ash, and lahars sped down the River Guali and River Lagunillas valleys as far as the River Magdalena. On 19 February 1845, about 1000 people perished when another lahar swept down the River Lagunillas. It destroyed Ambalema, 30km beyond the site where Armero was to grow up at the exit to the River Lagunillas canyon on the plain coated by this very lahar. Similar events were repeated in the same places in 1985 – except that the lahars were *less* extensive than in 1845.

Small earthquakes on 22 November 1984 first warned of the impending eruption, and 65 further earthquakes accompanied the initial minor phreatic explosion on 22 December. Sporadic earthquakes, increased fumaroles and columns of steam over 100m high, sometimes accompanied by fine ash and sulphurous fumes, marked mild, intermittent activity until August 1985.

Five days of regular earthquakes, repeated at hourly intervals, began on 6 September, which heralded a vigorous phreatic eruption at 13.30 on 11 September that continued for about 7 hours. Blocks fell 2km from the Arenas crater and 1 cm of ash blanketed the summit, melting enough ice to generate a lahar at 18.30 that swept 27km down the River Azufrado valley in an hour. Further earthquakes and phreatic eruptions, especially in the last week of September, scattered ash sometimes as far as Manizales. Three days of continuous earth-tremors began on 10 November, which indicated that the magma was approaching the summit.

At 15.06 on 13 November 1985 a phreatic eruption scattered rain-sodden ash up to 50km northeast of the crater. At Armero, ash began to fall at 17.00 but was replaced by torrential, rather muddy rain at 19.00. The first eruption of magma began at 21.08, when six andesitic-dacitic nuées ardentes at temperatures exceeding 900°C were accompanied by a small column of ash that had risen about 8km by 21.30. One-tenth of the summit ice-cap melted and generated several lahars (Fig. 6.4).

One lahar flowed westwards down the River Cauca, submerged Chinchina 70km away at 22.30, and killed 1100 people before evacuation could be completed. Other lahars sped down the River Guali and the upper River Lagunillas. But the lethal lahar formed about 20.20 in the upper River Azufrado valley, where a hot glowing-yellow mixture of ice, meltwater, rain, ash, pumice, thick rain-sodden older volcanic fragments and vegetation soon roared forwards like a pulsating mass of newly mixed concrete. The lahar followed the River Azufrado valley into the River Lagunillas, where it gained increased vigour by bursting a dam formed by a landslide in 1984.

Preceded by waters released from the dam, the lahar surged from the River Lagunillas canyon at a speed of 30 km an hour, and turned across the plain towards Armero with a front almost 40 m high. The lahar arrived in Armero, 60 km from its source, at 23.30 and swept away everything in its path: buildings, trees and people.

"There was total darkness", said a survivor. "A river of water came down the streets . . . dragging beds along . . . sweeping people away . . . The hotel [was] a three-storey building . . . built of cement and very sturdy . . . I saw [something] like foam coming down in the darkness . . . it was mud approaching the hotel . . . it crashed against the rear of the hotel and started crushing walls . . . and the entire building was destroyed and broken into pieces . . . the mud was hot but not burning . . . "

The scene that greeted rescuers on 14 November 1985 was of indescribable horror. People were dead and dying in the mud, trapped cows moaned, dogs ate mutilated bodies. About 25 000 people and 15 000 animals were killed; many survivors were injured by wrecked masonry; only 100 out of 5000 houses remained intact. Most of Armero had vanished beneath more than 5 m of mud. The cemetery was spared.

As Nevado del Ruiz had already been rumbling in a minor key for 51 weeks, the catastrophic eruption was not a complete surprise. But, although activity had increased in September, the volcano did not seem to be building up to a crescendo. Indeed, a scientific group visited the crater on 12 November and saw nothing exceptional. The eruption was small. Less than 0.0005km^3 of magma was ejected which melted between 0.01km and 0.02km^3 of water from one-tenth of the ice-cap. Together they formed surprisingly large lahars, but only totalling 0.06km^3 in volume – perhaps because of the torrential rain and the copious amounts of sodden surface materials available for mobilization. But, although such lahars had been widely – but not invariably – feared, no monitoring techniques had been installed that could pinpoint when they might be generated, or effectively warn those in danger once they had set off.

When the eruption started, Colombia had no monitoring equipment, and lacked both funds to buy it and qualified volcanologists to use it. On 9 March, a visiting seismologist suggested that seismographs be installed, a hazard map drawn up and refuges designated. No refuges were apparently ever designated in the northeastern valleys. The first seismographs imported to register the earthquakes caused by rising magma were installed on 16 July, but they often worked badly because they arrived in poor condition. When they were replaced, the seismograms were often sent, unassessed, to Bogotá, whence the details were released only infrequently. But in any case the network was probably installed too late, and too sparsely, to predict eruptions accurately enough. Work finally began on the hazard map on 20 September. The first edition, published in the press on 9 October, quite accurately forecast the areas later damaged by lahars.

Throughout 1985 experts usually representing foreign agencies visited Nevado del Ruiz. Even after the eruption on 11 September, they came to surprisingly different conclusions about how the volcano might behave. One, for example, declared that only the summit area would be affected and that there was no danger from lahars; others, like the expert Italian team on 22 October, strongly urged that priority be immediately given to establishing an efficient system of communicating lahar alerts and designating places of refuge for the threatened populations. After months of scepticism, the Colombian Institute of Geology and Mines stressed on 8 October that an eruption would certainly cause lahars that would put several towns, including Armero and Chinchina, in grave danger.

But ordinary people were still confused because, as outsiders, they had no means of assessing which "expert" was right. They were neither reassured nor further enlightened in early October when an archbishop condemned those who, he said, were spreading alarming volcanic terrorism. Mammon also intervened when other authoritative non-experts complained that the volcanic risk map would reduce property prices. Thus, the mayor of Armero could well complain on 8 October that the citizens had lost all confidence in the truth of the information given and had put their fate in God's hands.

The emergency was also handled differently in different provinces. In Manizales, in Caldas province in the west, an *ad hoc* volcanic risk committee was soon set up to gather and disseminate information, hosting Swiss and Italian expert teams, as well as preparing contingency disaster plans. Thus, evacuation, which could only be partial with less than two hours' warning, saved lives when the lahar struck Chinchina. No such co-ordination appears to have been established in Tolima province in the east, where Armero and other towns in the River Magdalena valley were generally acknowledged to be most vulnerable to lahars. On 7 October, the Institute of Geology and Mines claimed that Armero would be evacuated in two hours if a lahar came, but it is unclear how it could have been implemented. Responsibility was deliberately devolved into the provincial and municipal authorities on 16 September. But they had restricted powers, limited resources, conflicting information and little methodological basis for planning for the emergency effectively, especially on the east of the volcano.

Foreign agencies were also largely unable to act until October, and, even then, protocol limited them to recommendations. It is difficult to interfere in a sovereign state's eruption. For instance, a letter from UNESCO encouraging the government to ask it for help was lost in the Colombian bureaucracy for two months and the measures urged by the Italian volcanologists were largely ignored. On the other hand, one expert agency haggled over costs when asked for help.

There was, therefore, a desperate need for authoritative, well informed guidance and contingency planning which only a committed central government could provide. One problem for the administration was to decide which towns to evacuate and when. Months of initial scepticism were followed by procrastination because no-one could, or would, take the responsibility for evacuating a large population, or for designating places of refuge – even if proper contingency plans had been made. Because nobody could predict exactly when a lahar-forming eruption would take place, the administration waited, fearing false alarms, until the last possible moment before taking any action. There was certainly a danger of civil unrest if evacuation proved too early or unnecessary. Indeed, the Colombian government had other problems: a minor insurrection, dictated by non-volcanic matters, was put down with bloodshed on 6 November in Bogotá. Civil defence planning was therefore only limited and unco-ordinated.

Thus, on 13 November 1985, Armero was taken unawares. At 21.45 a Red Cross official in the provincial capital radioed to urge the evacuation of Armero but, although individuals were warned, the order was not systematically communicated to the town. One fireman tried to alert the people by blowing a whistle in the streets, but, by then, it was bedtime, it was raining heavily, and an important soccer match was being televised. Most people refused to move, preferring to believe the reassuring messages that the priest and the mayor had broadcast earlier on the local radio. Radio Armero then played cheery music until the lahar arrived.

Hindsight fosters criticism, but also reveals an important extenuating circumstance. The lethal lahars were generated by the very first eruption of hot magma and they struck Armero and Chinchina within about two hours. Thus, it proved vital to be able to predict exactly when the magma would arrive at the surface, which is notoriously difficult in subduction zone volcanoes. It was an impossible task with the monitoring methods employed, and especially with the lack of communication between the experts and the people, and in the absence of real disaster planning.

Mount St Helens, 18 May 1980

Mount St Helens rises in Washington State in the Cascade Range that graces the American Northwest from California to British Columbia. Along with Mount Adams and Mount Hood, it stands guard over the Columbia River valley and, before the eruption in 1980, reached a height of 2950m (Fig. 6.5). Mount St Helens had long been famous for its ethereal beauty amongst some of the world's most distinguished mountains and, since 1980, has become one of the most notorious volcanoes on Earth. But the beauty of Mount St Helens was masking a recent past of greater violence than any of its Cascade companions. Thus, when American geologists published hazard studies of the Cascade volcanoes in 1975, they forecast that Mount St Helens could well erupt again before the end of the century.

Since 1792, when Vancouver discovered the volcano and named it after a British diplomat, Mount St Helens had erupted in about 1800, in 1831 and in 1836, and sporadically between 1842 and 1857. Although the short historical record provided only scanty information, geological studies showed that Mount St Helens usually erupts with great violence. Nine such phases have punctuated the 30000 year history of the volcano and most of its cone has grown up during the past 2500 years. The Native Americans, who knew the Cascade Range long before Vancouver, must also have recognized the true nature of Mount St Helens because they called it "One From Whom Smoke Comes", or "Fire-mountain".

Thus, the eruption of Mount St Helens was not a complete surprise to a few well informed Earth scientists. What was surprising was the intensity and manner of its

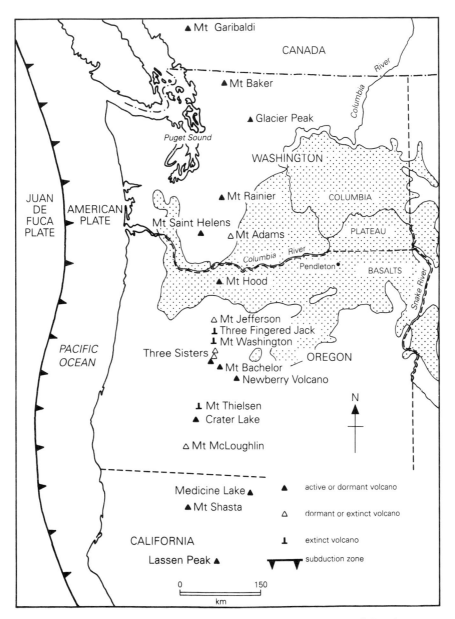

Figure 6.5 The major volcanoes of the Cascade Range, USA and Canada.

advent. Fortunately, two months elapsed between the first inkling of reawakening and the great eruption – enough time for the great technological resources of the United States to be harnessed and register the volcano's every caprice.

The first suggestion that an eruption might be imminent came with a magnitude

4.0 earthquake below the volcano on 20 March 1980. Other earthquakes in the following week all had a shallow focus and were centred below the northern flank of Mount St Helens. The eruption duly began on 27 March, when an explosion blasted an 80 m crater in the 350-year-old summit dome. Similar eruptions, accompanied by continual earthquakes caused by the rising magma, followed until 22 April, and enlarged the crater to an elliptical hole that was soon 500 m across and 200 m deep. All these eruptions were phreatic, caused when water from melting summit ice seeped down and was suddenly heated to steam when it met the rising magma.

Measurements also revealed that the northern flank of the volcano had started to swell outwards as early as 27 March. This sinister bulge was soon expanding at 1.5 m a day, and the focus of the tremors rose above the base of the volcano to a mere 2000 m below the summit, showing that the magma had invaded the cone. By mid-April, the bulge was 1600 m long, 800 m wide and swelled some 100 m outwards.

On 7 May an eruption formed an ash and steam column that rose 4 km into the air. By this time, 8000 tremors had been registered and, more significantly, they were getting stronger. In the next week the bulge expanded to 2 km across and swelled out by some 150 m. On 12 May a magnitude 5.0 earthquake caused a small debris avalanche on the northern slope. Its significance was not appreciated at the time, but there was not long to wait.

Sunday 18 May 1980 at Mount St Helens was a beautiful morning, ideal for photography. With great sang-froid, a lucky and skilled photographer recorded the events from what proved to be a safe distance. An earthquake, a debris avalanche, a gigantic blast and a violent eruption of pulverized fragments of lava, succeeded each other in less than a minute, decapitated the cone, devastated a vast area and made Mount St Helens immortal (Fig. 6.6).

At 08.32, a magnitude 5.0 earthquake occurred at a depth of only 1600 m beneath the northern flank of the volcano. In the next ten seconds the bulge rippled like jelly before giving way in an enormous debris avalanche that swept 3 km^3 of rock down slope at a speed of 250 km an hour (Fig. 6.7). It scoured bare the northern slopes of the volcano, surged across Spirit Lake, and over a 360 m ridge and down the north Toutle River for 22 km. Eight thousand million tonnes of rock were removed from the cone.

The enormous confining pressure exerted by the volcano on the rising magma was almost instantaneously released. At once a great blast exploded sideways from the northern flank which had been gashed open by the debris avalanche. The magma was probably laid bare. The hot mixture of rock, pulverized magma, gas, steam and ice broke loose, and, travelling at almost the speed of sound, soon overtook the debris avalanche and spread well beyond it. The blast lasted only 1½ minutes, but devastated an area of 600 km^2, in which millions of noble conifers were blown down like matchsticks. David Johnston, the geologist on duty that morning at his post 10 km northwest of Mount St Helens, just had time to contact his base at Vancouver, Washington, and shout over his radio: "Vancouver, Vancouver, this is it!". Then he died.

Figure 6.6 Mount St Helens, with the dome in the debris avalanche caldera, and blasted tree-trunks in the foreground.

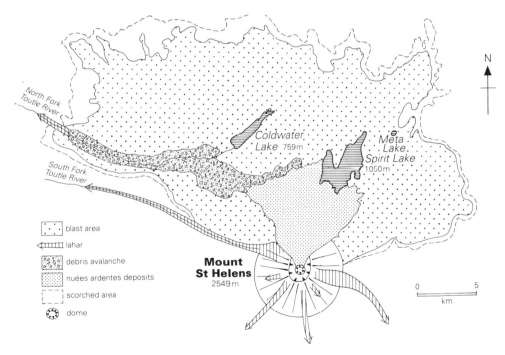

Figure 6.7 Map of Mount St Helens after the eruption of 18 May 1980.

A few seconds later, just before 08.33, a vertical eruption formed an enormous billowing, columnar cloud of dark ash. It surged up to reach 25 km into the stratosphere within the next 15 minutes. This Plinian eruption lasted until 17.30 and formed a typical pine-tree-shaped cloud that spread ash and darkness far into eastern Washington. About 60 000 km² were covered with 540 million tonnes of volcanic dust. Carried on eastwards by high-altitude winds, the cloud reached New York on 21 May, and made its first circuit of the globe in the next two weeks.

Meanwhile, as the column continued to rush upwards on 18 May, nuées ardentes repeatedly sped down the northern slopes at speeds of up to 300 km an hour. They covered the debris avalanche, boiled streams and lakes and blanketed the valleys well beyond the northern base of the volcano.

At the same time on 18 May, lahars were formed that destroyed practically everything in the valleys as much as 40 km away from the cone (Fig. 6.8). They were composed of hot ash mixed with water from lakes and streams and especially from melting ice and snow, as well as uprooted trees, fragments of buildings and the remains of previous eruptions. The lahars moved more rapidly than on Nevado del Ruiz, attaining speeds of 150 km an hour and resembled floods of boiling porridge 20 m deep. The biggest lahar swept into the north Toutle River at 14.00, ripped up twisted bridges, wrapped a logging truck around a tree and had enough momentum to flow over hills 90 m high.

By the evening of 18 May 1980, Mount St Helens had lost 440 m from its summit and had been gutted on its northern side by a great horseshoe-shaped hollow which was 3.2 km long by 1.6 km wide and 700 m deep (Fig. 6.6). The volcano had dis-

Figure 6.8 Edge of a lahar from 1980 eruption, Mount St Helens.

placed $3 km^3$ of debris avalanche and erupted at least $1 km^3$ of magma. In its nine hours of activity on 18 May, Mount St Helens had released energy equal to the explosion of 27 000 Hiroshima atomic bombs – at a rate of almost one for every second of the eruption. The cost of the damage was estimated at US$1000 million. Fifty-seven people were killed. Superb coniferous forests were destroyed, but, towards the fringes of the devastated area, at Meta Lake for example, tree saplings and hibernating beavers, beneath a protective cover of deep snow, as well as frogs and trout in frozen lakes, often survived. They were in a kind of "baked Alaska", where the hot blast passed over too quickly to melt the ice and snow completely.

Mount St Helens remained active after 18 May and several spectacular eruptions of ash and nuées ardentes occurred, for instance, on 25 May, 12 June, and 7 August. The last major display occurred on 16 October 1980. Subsequent eruptions were never strong enough to destroy the dome that began to form over the vent in the caldera. Mount St Helens then entered upon a more constructive phase that lasted until 1986. Few changes have taken place since, but it might be expected that small eruptions will fill the caldera in the next few centuries. The stage will then be set for another great outburst.

Difficult though it may be, it is important to retain a sense of proportion about Mount St Helens. It has been studied in extraordinary detail and has given new scientific insights into the effect of earthquakes, blasts and debris avalanches on volcanoes. But this was not the greatest eruption of the century. Santa Maria (1902), Katmaï (1912), Bezymianny (1956) and Pinatubo (1991), for example, were more powerful – and several have killed more than 57 people. On the other hand, Mount St Helens has perhaps made the greatest contribution to our understanding of volcanoes since the eruption of Montagne Pelée in 1902.

Montagne Pelée, 8 May 1902

The French West Indian island of Martinique lies in the centre of the Lesser Antilles volcanic arc, which was generated when the North American Plate was subducted beneath the Caribbean Plate. The Lesser Antilles stretch like a rosary in a curve 600 km in length between Puerto Rico in the north and Grenada off the coast of South America (Fig. 5.3). They are composed of about 20 major volcanic islands, amongst which 18 may be considered active. Volcanic activity in Martinique has shifted northwestwards as the arc has developed. The remains from the oldest, long-extinct eruptions thus now form low hills in the southeast. The middle of the island is occupied by the largely basaltic shield of Morne Jacob, formed some 5 million years ago, and by the ten andesitic domes comprising the Pitons du Carbet, which erupted with some violence about 800 000 years ago. Both these centres are almost certainly extinct: the torch has been passed to Montagne Pelée.

Montagne Pelée is all the more sinister because it presides over a tranquil and beautiful palm-fringed bay of superb proportions (Fig. 6.9). Indeed, before visiting St Pierre, it is hard to believe that the finest town in the West Indies was destroyed

Figure 6.9   Montagne Pelée from St Pierre.

in two minutes by a volcano that has hardly raised a fumarole in the past 60 years. The volcano rises to 1397 m and covers most of the northwestern peninsula of Martinique (Fig. 6.10). Montagne Pelée began life about half a million years ago as a submarine volcano, which quickly accumulated above sea level. The bulk of the edifice, though, is less than 35 000 years old, and was piled up by brief, alternating phases of Pelean and Plinian eruptions. Its lavas were usually viscous enough to form domes when their gas content was depleted, but Montagne Pelée expelled many layers of fragments during its violent outbursts. Lava-flows have been uncommon, and none has been emitted during historic time. Last century, the summit was marked by a caldera, 1500 by 800 m in size, which was probably formed about 2000 years ago. It was 100 m deep, and contained a small lake that only filled in the rainy season and thus was known as the Etang Sec. The edge of the caldera was notched on the southwest, facing St Pierre, by the headwaters of the Rivière Blanche. An extreme pessimist might have feared that an eruption from the caldera would thereby be directed straight towards St Pierre, 7 km away to the south.

At the turn of the century, however, the historical records gave no hint that a major eruption at Montagne Pelée was to be feared. Since Martinique was first settled in 1635, only two weak eruptions had occurred, in 1792 and 1851, which had done little more than excite mild curiosity amongst the inhabitants. However, fumarole activity resumed at the summit in 1889, and had reached such intensity by February 1902 that the hydrogen sulphide exhaled tarnished the household silver in St Pierre. In March and April, nauseating smells were common in Le Prêcheur and other villages on the north coast. No-one realized that these apparently trivial signals were heralding the most lethal eruption of the century . . .

In 1902, St Pierre was the "pearl of the West Indies", a town with a population of over 26 000, proud of its electric light, bustling port, rum distilleries, flourishing chamber of commerce, and its theatre that was an exact copy of the Grand Theatre in Bordeaux. On 24 April 1902, a phreatic explosion expelled ash from Montagne Pelée for the first time (Fig. 6.10). The following morning, Le Prêcheur was pow- dered with just enough fine grey dust to cause a certain unease in the little port. However, calm returned on the next day, 26 April, and a party climbed the moun- tain to see what was afoot. The intrigued and insouciant visitors to the crest saw a steaming, leaden-looking lake in the usually dry Etang Sec, with blackened vegeta- tion around it, and a new cone of cinders, 10m high on its eastern shore. The sum- mit remained relatively calm until 2 May, with only small phreatic ash eruptions and a few earth-tremors, felt as far away as St Pierre. On 27 April the first round of the legislative elections was held, which saw the opposition candidate in the lead. (The next round was due a fortnight later on 11 May.) After phreatic eruptions expelled more dense, dark grey columns of ash and steam during the afternoon of 2 May,

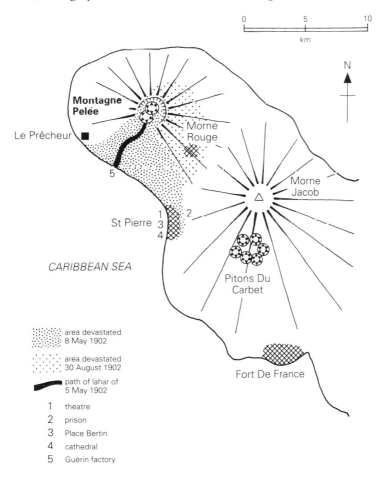

Figure 6.10   Map of Montagne Pelée and northwestern Martinique, and the eruptions of 1902.

which were accompanied by detonations like distant gunfire, ash fell on St Pierre for the first time at 23.00. The people living on the flanks of Montagne Pelée began their exodus; and there was a rush to confession that night at Morne Rouge at the southeastern foot of the mountain. On 3 May, the opposition candidate in the elections called for the evacuation of St Pierre. The authorities thus emphasized the need for calm. The eruption was drawn into the electioneering as extremely fine ash continued to rain down on St Pierre, and 5 cm soon accumulated like snow on the rooftops. Meanwhile Le Prêcheur was in total darkness. On the morning of 3 May, Governor Mouttet arrived in St Pierre from the capital, Fort de France, and journeyed on to Le Prêcheur to improve morale. The schoolchildren were given a holiday because the ash was now so thick. Birds could not fly; and "the ash had turned the blacks in St Pierre white". Although the intensity of the activity abated on 4 May, noisy eruptions during the following night scattered thick ash all over the north of the island and the Rivière Blanche, draining southwestwards from the volcano, began to flood.

At 12.45 on 5 May, the volcano suddenly expelled the waters of the Etang Sec through the notch in the rim of the caldera into the Rivière Blanche, and formed a lahar that swamped the Guérin factory at its mouth in 6 m of mud and 23 workers became the first victims of the eruption. In an another attempt to reassure the population, Governor Mouttet paid a second visit to St Pierre on 5 May, and stayed overnight before returning to Fort de France. It was widely held amongst the more informed people in the city that the Rivière Blanche valley would safely channel all the emissions from Montagne Pelée directly into the sea and that St Pierre itself was in no great danger. Hundreds of people had, however, already moved to the relative safety of Fort de France, especially if they could secure accommodation in the capital. The maintenance of sang-froid, however, was not helped when the electricity supply to St Pierre failed on 5 May. By that evening, too, all the Pelean streams were in flood, fed by the groundwater heated by the upwelling magma. Throughout 6 May, lightning continually flashed about the summit as fine ash fell persistently over St Pierre. The electricity supply had not been restored, but a mocking glow spread about the mountain as incandescent lava exuded from the vent. The magma had reached the surface. The mayor of St Pierre, Monsieur Fouché, tried to reassure the people by issuing a poster in which he called for everybody to resume their normal activities and not let themselves succumb to panic. Anxiety in St Pierre was, however, increased by the numbers of alarmed refugees still entering the city from villages farther north. On 7 May, "Les Colonies", the local newspaper, reported the reassuring conclusions of the *ad hoc* investigative scientific committee. One of its members, Monsieur Landes, a science teacher at the lycée, affirmed that "Montagne Pelée presents no more danger to St Pierre than Vesuvius does to Naples". That morning, however, Montagne Pelée had already expelled its first weak nuées ardentes that were too small, however, to reach the foot of the mountain. On the evening of 7 May, Governor Mouttet and his wife returned once more to St Pierre to spend the night in the city and thereby set an example of calm to the anxious people. A thunderstorm of awesome violence played about the summit all night.

Muddy torrents flooded all the villages on the north coast. At midnight, the chapel in Morne Rouge was filled by terrified people asking for the last rites.

Thursday 8 May was Ascension Day. The morning was calm, although a cloud of ash covered the sky over St Pierre. The governor, his wife and a small party of officials set off by boat for Le Prêcheur. They were never seen again. "I was telephoning a friend in St Pierre", a businessman from Fort de France said later, "He had just finished a sentence when I heard a dreadful scream, and then another much weaker groan, like a stifled death rattle. Then silence."

At 08.02 there was a blinding flash, and then a sound like the roar of thousands of cannons. A directed, low-angle blast, followed by a nuée ardente, left the summit by the little notch on the southwestern rim of the caldera, swept down the Rivière Blanche valley, rushed southwards beyond it at speeds estimated between 200 km and 500 km per hour, and at temperatures ranging between 200°C and 450°C, devastated an area of 58 km^2, and overwhelmed St Pierre in less than two minutes. The blast and then the incandescent cloud of swirling ash, steam, and an emulsion of lava blocks, boulders, pumice, toxic gases, trees and masonry hurtled down the streets, into the houses, into the cathedral crowded with worshippers, and killed more than 28 000 people in the space of two fatal breaths. Several more unfortunate citizens, including Monsieur Landes, whose home was on the very edge of the area covered by the nuée ardente, survived with badly burnt lungs for over an hour. The lighthouse in the Place Bertin was snapped off at its base. The boats in the port were wrecked and overturned. Only the main arch and hall pavement of the lycée remained. Nearly every wall athwart the line of the blast was thrown down. Often only rubble, stone bollards and pavements marked the pattern of the streets. Little but the towers of the cathedral still stood. Not much more than the front steps of the theatre and the chamber of commerce could be seen. The rum distilleries were surrealistic piles of twisted metal. A few crewmen were still alive on ships moored off-shore beyond the limits of the cloud. Within the area covered by the nuée ardente only a shoemaker, Léon Compère-Léandre, survived on the extreme outskirts of the blasted zone and one famous prisoner, Louis-Auguste Sylbaris, incarcerated the previous evening for disorderly conduct, lived on, badly burned and half mad with thirst and anxiety, to be rescued from his prison in a hole alongside the theatre on 12 May (Fig. 6.11). He told his rescuers: "All of a sudden there was a terrifying noise. Everybody was screaming 'Help! Help! I'm burning! I'm dying!!' Five minutes later, nobody was crying out any more – except me."

Puzzled by the sudden extinction of all communications from St Pierre, several citizens of Fort de France set off by boat to see what had happened. They could scarcely comprehend the scene as they rounded the cape at Le Carbet. Until that day no human eye had witnessed such urban devastation, not even in the great earthquake of Lisbon in 1755. It still remains the largest volcanic death-toll of the twentieth century (Fig. 6.12). The northern part of the city was flattened, but a few walls remained in the south. Parts of the city were ablaze. Several boats had sunk. Most of the dead were blackened and many were buried beneath the layer of ash that enshrouded St Pierre. They had been asphyxiated and burned, and many had fallen

Figure 6.11   The prison of Sylbaris, St Pierre.

Figure 6.12   The ruins of St Pierre, May 1902.

facing southwards, thrown down by the blast. There was a great silence of the Apocalypse everywhere. Nobody, as far as they could see, was left in St Pierre to vote.

On 20 May another blast and nuée ardente threw down most of what still remained upright in St Pierre; others occurred with decreasing intensity on 26 May, 6 June and 9 July. On 30 August another nuée ardente surged southeastwards and devastated Morne Rouge, killing a further 1000 people. This proved to be the last major outburst, although eruptions continued, on a smaller scale for more than a year. Activity then was gradually stifled as an andesitic dome grew above the vent and continually thrust out high, nearly solid spines. One such spine, the sinister "Needle of Pelée", was extruded to a height of 310 m in June 1903, but many others reached 40–50 m before they shattered. Dome building went on slowly until October 1905. The ensuing period of calm lasted until the Autumn of 1929, when Montagne Pelée erupted again, but less violently than in 1902. Another dome was formed beside, and partly destroyed, the first, so that they both nestle in the summit caldera like two enormous cuckoo's eggs. The tropical heat and 8 m of annual rainfall has encouraged the growth of a thick mat of vegetation that masks the clefts and makes walking on the domes hazardous. These clefts have been the only threats since the latest eruption ended in 1932, for the summit has shown no subsequent sign of activity. But Montagne Pelée is not extinct; it is only biding its time.

The Minoan eruption of Santorini

Santorini in Greece is the most spectacular island in the Aegean Sea and one of the most impressive volcanic sites in the world. It was the scene of the largest eruption in Europe since the Ice Age (Fig. 6.13). The outer slopes of Santorini are gentle, but its central area has been gutted by a steep-walled caldera, flooded by the Aegean Sea, which now stretches 11.5 km from north to south and 8 km from east to west (Fig. 6.14). The caldera is fringed by fragments of the partly destroyed island. Therasia and Aspronisi lie to the west, and Thera forms the largest survivor, dominating the blue arena with sheer cliffs, often 400 m high, composed of brown, red or black lava-flows in the north and buff layers of pumice in the south (Fig. 6.15). The Kameni ("Burnt") Islands have erupted in the midst of the caldera since 197 BC.

The Bronze Age, or "Minoan", eruption gave Santorini both its notoriety and its spectacular landscape. The eruption buried a city, excavated at Akrotiri since 1967, which revealed some of the most exciting archaeological finds of the century (Fig. 6.16). Akrotiri was a Minoan colony, named after the King Minos, who ruled from the palace of Knossos in Crete. This highly developed civilization, lasting a millennium, suddenly ended in the Bronze Age, and many have thought that the eruption of Santorini sounded its death knell. But there are problems with this spectacular thesis – problems that drew together the disciplines of geology, geography and archaeology, and fuelled controversy amongst their exponents, notably when doubt was cast on the real date of the eruption, on the size of the caldera formed, and on its true rôle in the demise of Minoan civilization.

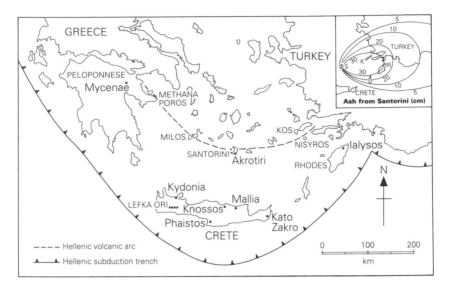

Figure 6.13 The Minoan Aegean, Santorini and the Hellenic volcanic arc, with the distribution of ash from the Minoan eruption.

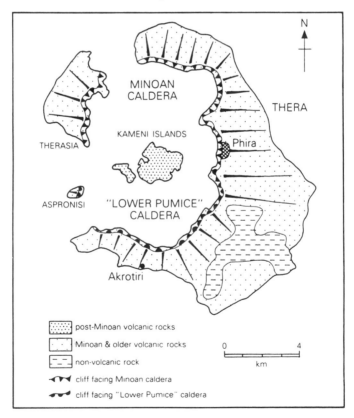

Figure 6.14 Map of Santorini, Greece.

Figure 6.15 The Minoan caldera edge from Phira, Santorini.

Figure 6.16 The ruins of Akrotiri, Santorini.

The Minoan eruption

During the past 200000 years, Santorini experienced at least a dozen violent eruptions, accompanied by several caldera collapses. One of the most important developed the "Lower Pumice" caldera about 100000 years ago. Flooded by the sea,
6km across 280m deep and open to the southwest, it now occupies the southern
part of the present caldera. Thereafter, activity continued, especially to the north,
making an undulating landscape of several low volcanoes, about 300–400m high,
that almost enclosed the "Lower Pumice" caldera. Thus, the southern part of the
caldera already existed before the Minoan eruption, and Santorini probably formed
a volcanic field, resembling the present island of Milos, rather than a great strato-
volcano as was long believed. In fact, the Minoan landscape of Santorini may well
be represented both in the left and right hand scenes of the "naval wall-painting"
from the West House at Akrotiri, which is now in the Archaeological Museum in
Athens.

The inhabitants of Akrotiri were warned of the impending eruption by premonitory earth-tremors, which were powerful enough to damage sturdy, stone-built,
two- and three-storey buildings in the city. These were being repaired when a preliminary phreatic eruption began. Forewarned, the Akrotirians abandoned their
city, and thus no human bones have been found there, but it is doubtful if *any* inhabitants remaining on Santorini managed to survive the ensuing cataclysm.

The Minoan eruption occurred in four relentless phases and lasted about four
days. The first phase, lasting about eight hours, was marked by two Plinian columns,
rising perhaps 36km, which distributed hot, coarse "rose" rhyodacitic pumice that
reaches 6m in thickness.

The second phase gradually became more hydrovolcanic as magma erupted
violently within the flooded "Lower Pumice" caldera. Ash and pumice exploded
vertically and surged sideways and knocked down walls in Akrotiri. This second
phase probably lasted about a day.

The third phase was also hydrovolcanic and was responsible for more than half
the materials expelled by the Minoan eruption, forming massive chaotic layers of
white rhyodacitic pumice, lapilli and ash, replete with blocks of old lava and breadcrust bombs. These beds reach up to 55m in thickness and are well displayed in the
pumice quarry south of Phira. It was then that the Minoan caldera began to founder
as lahars and ashflows surged down all the outer slopes of Santorini. This phase may
also have lasted a day.

The final phase, which may have lasted only several hours, saw the expulsion of
ochre-coloured pumice and ashflows, and the creation of the Minoan caldera. Its
collapse began in the west and progressed eastwards – like a key pressed down on a
piano keyboard – until a caldera some 8km wide and 9km long had foundered in
the northern part of the island and joined the "Lower Pumice" caldera where the
Kameni Islands were later to erupt from the fractures created. Its confining rim still
has steep walls, which, in places, rise 700m from its flooded base lying 380m below
sea level. About 36km^3 of pumice was ejected during these four days.

The date of the Minoan eruption

The Minoan civilization was centred on Knossos, with secondary palaces at Mallia, Phaistos and Kato Zakro, and a scattering of settlements throughout the Aegean area. Although distinctions are blurred, and opinions differ, archaeologists have subdivided its final phases into Late Minoan IA, between about 1550 BC and 1500 BC, Late Minoan IB, from about 1500 BC to 1450 BC, and Late Minoan II, from about 1450 BC to 1400 BC. The divisions are based primarily on changing pottery styles and dateable Egyptian finds in the Minoan sites. All the Cretan palaces, except Knossos, were suddenly destroyed at the close of Late Minoan IB. Knossos survived until the end of Late Minoan II, but under the occupation of Mycenaean Greeks, who had apparently profited from the mysterious calamity to establish their hegemony over the Aegean area.

Conquest and earthquakes were first invoked to explain the catastrophe, but it was then proposed that eruptions of Santorini had generated ashfalls, earthquakes and, especially, a tsunami which had brought Minoan civilization to its knees. The similar, but smaller eruption of Krakatau in 1883 provided a guide to the possible course of events – settlements crumbled by earthquakes; countryside blanketed by ash, killing crops and animals; tsunamis, formed by caldera collapse, crashing onto the northern shores of Crete at speeds of 250 km an hour; with many survivors of this awesome sequence dying in the ensuing famine. Even this most sophisticated of European Bronze Age civilizations would surely have tottered and collapsed in the face of such an onslaught . . .

Excavations at Akrotiri from 1967 onwards appeared at first to confirm this view, but it soon became clear that the eruption had buried the city at the end of Late Minoan IA – and not in Late Minoan IB when the Cretan palaces had been destroyed. Thus, the eruption of Santorini could not have destroyed the Cretan palaces – unless either the Minoan eruption had lasted for 50 years or the chronology of the Aegean area (and that of Egypt to which it was linked) was defective. Although scientists would normally be delighted to be able to date a Bronze Age event to within 50 years, the timing of the Minoan eruption had to be established with even greater precision. The problem became even more crucial when geological studies increasingly indicated that the eruption had lasted only a few days, and certainly not 50 years.

The quest for the exact date of the eruption was extended beyond the realms of archaeological finds when further evidence was adduced from radiocarbon dating, acidity concentrations in polar ice-caps, and dendrochronology based on tree-ring growth. This research produced as many problems as solutions – not least by offering different dates altogether for the eruption! Radiocarbon dates were obtained from short-lived shrubs killed on Santorini by the eruption. They ranged disconcertingly from 1675 to 1525 BC, although statistical tests pinpointed 1629–1622 BC as the most likely dates. Nevertheless, they were a century before those indicated by the archaeological record. However, the wide range of these dates made them very suspect unless they could receive independent corroboration.

Acid aerosols formed after major eruptions have often been preserved in layers in polar ice-caps, but it is hard to link them to specific volcanoes. Cores taken from the Greenland ice-cap showed aerosol acidity concentrations, dated to about 1645 BC, and seemed to offer some corroboration of the radiocarbon dates. However, the link to Santorini was weak and wholly circumstantial, and was apparently made because it was the only known eruption thought capable of causing such an acid concentration at that time. In fact, the ice-cores revealed between six and twelve times more volcanic aerosol fall-out than that now calculated for the Minoan eruption and therefore another volcano must have expelled them.

Meanwhile, dendrochronological studies entered the argument. Thin tree-rings indicate poor annual growth which is often attributed to cold weather (which is not necessarily caused by great eruptions). Thin tree-rings in bristlecone pines in the southwestern United States were attributed to a cold spell about 1627 BC, provoked by a dust veil that could have spread from Santorini. Ancient oaks found in bogs in Northern Ireland also showed small tree-ring growth, caused by a very cold spell about 1628 BC. But, even if this poor growth were really caused by an eruption, no firm link could be forged with Santorini. Many explosive volcanoes, including a dozen in Kamchatka and the Aleutian arc alone, were better situated and just as likely as Santorini to be the culprit. But, unfortunately, few have been studied in detail. Who knows (yet) what each one was doing between 1650 BC and 1600 BC? However, recent studies have revealed that, in Alaska alone, Aniakchak, Veniaminof and Hayes (recognized as a volcano only in 1975) all experienced violent eruptions between 3400 and 3800 years ago. Thus, although the Minoan eruption could have occurred about 1645 BC or about 1627 BC, these links are not proven. Moreover, if they were adopted, the Bronze Age chronology throughout the eastern Mediterranean area would have to be thoroughly revised. In any case, whichever date is adopted for the eruption, it still fails to solve the problem presented by the fact that Akrotiri was apparently destroyed 50 years before the Cretan palaces.

The effect of the eruption on Minoan civilization

If the eruption of Santorini caused the demise of Minoan civilization, cause and effect must have followed each other rapidly and in a clearly related sequence. Just as doubt can be cast on the timing of the eruption, so its effects on Crete may be called into question. The collapse of the Minoan civilization has been attributed to a military takeover by Mycenaeans, to extensive earthquakes, to volcanic ashfalls, to a tsunami, and to various combinations of these scourges.

Santorini is irrelevant to a purely political or military Mycenaean conquest of a weaker Minoan state, but it is hard to imagine why they would have so severely damaged the palaces. The eruption of Santorini could not have been responsible for widespread earthquakes because those related to rising magma would have affected only Santorini and its immediate vicinity, leaving Crete, 125 km away, unscathed. Any severe Cretan earthquakes – at any date – must have been related to crustal movements caused by the subduction of the African Plate beneath the Aegean. In any case, when earthquakes had previously destroyed Cretan palaces about 250 years before, they were simply rebuilt in even greater splendour.

It is hard to evaluate the effects that copious pumice and ash falls might have had on a Bronze Age civilization, but, although famine, death, emigration and civil disruption may well have ensued, they could hardly have toppled the palaces. Also, newly discovered deposits now indicate that most of the Santorini ash was blown over Turkey, and that Crete probably stood on the scarcely affected fringes of the cover, where ash did not exceed 5 cm in depth (Fig. 6.13). In addition, a 10 cm-thick pumice layer from Santorini that fell onto the Minoan colony at Ialysos in Rhodes did not even disrupt the continuity of the settlement, let alone annihilate it. This pumice also lies *beneath* the Late Minoan IB pottery layer, which offers independent evidence that the eruption, as at Akrotiri, occurred in Late Minoan IA, before the Cretan palaces were destroyed. A possible tsunami is thus the only remaining volcanic element that Santorini might have sent to destroy Minoan civilization. Speculation about its effects has been fuelled by legitimate comparisons with that generated by Krakatau in 1883, which, however, did not turn corners very easily. The collapse of the Minoan caldera could well have generated a tsunami, strong enough to topple even the sturdily built palaces of Knossos and Mallia near the north coast of Crete. But Phaistos, near the south coast, on a hill 200 m high, and protected from the north by the Lefka Ori, the highest mountain range in Crete, could not have been touched by a tsunami, although it was destroyed at the same time as the others. The direction taken by the tsunami presents another difficulty. The Santorini caldera is not breached in the south, facing Crete, but in the west and northwest. The tsunami would thus have travelled mainly towards the Peloponnese, making Mycenae its probable victim, rather than its indirect beneficiary (Fig. 6.13). At all events, the tsunami would have been generated when the caldera collapsed (and would have reached Crete 30 minutes later) at the close of Late Minoan IA, and could not have destroyed the Cretan palaces 50 years afterwards in Late Minoan IB. Thus, any tsunami probably went the wrong way at the wrong time – with few consequences on Crete.

It thus seems hard to blame the Minoan eruption of Santorini for the collapse of Minoan civilization. The geological consensus indicates that the eruption took place within about four days, which archaeological evidence places at the end of Late Minoan IA, about 1500 BC, whereas all the Cretan palaces except Knossos were destroyed at the end of Late Minoan IB. There is no firm evidence to link Santorini with acidity peaks in the Greenland ice-cap, or to the weak, frost-damaged growth in American bristlecone pines and Irish bog oaks that can be dated to about 1645 BC or 1627 BC. (The radiocarbon dates from Akrotiri, however, seem too unreliable to be taken into account.) But none of these dates offers any help with the time gap between the eruption and the destruction of the Minoan palaces. Thus, it is still probably more prudent to assume that the eruption took place about 1500 BC. The Cretans were no doubt greatly impressed by the Plinian column rising 36 km above Santorini and by the loudest noise in the Bronze Age, which would have been easily audible at the sites of Glasgow and Gibraltar. The evidence from geological, morphological and archaeological research, therefore, now apparently absolves the eruption from having played a significant rôle in the collapse of the Minoan civilization. What seemed an exciting correlation 25 years ago now seems much less certain. Moreover, as the recent changes in eastern Europe demonstrate, sudden events do not necessarily have sudden causes. Perhaps widespread violent earthquakes came when the Minoan civilization was already decaying and provided the *coup de grâce* from which the burgeoning power of Mycenae amply profited. Nevertheless, although it is deprived of an impact on one of Europe's most refined civilizations, the Minoan eruption of Santorini still remains one of the great volcanic events in recent geological history.

Capelinhos, 29 September 1957

The island of Faial in the Azores is noted for its luxuriant vegetation, but recent eruptions on its western peninsula have created a new, harsh, mineral world that juts out defiantly into the Atlantic Ocean (Fig. 6.17). The latest eruption began in September 1957 in the Atlantic Ocean off the Capelinhos lighthouse on the Costa da Nau, the westernmost bastion of the peninsula (Fig. 6.18). It occurred near the Ilheus dos Capelinhos, islets which were themselves remnants of a former volcano almost obliterated by marine erosion. The growth of Capelinhos volcano was thus only regaining ground previously lost to the Atlantic waves. Capelinhos first ex-

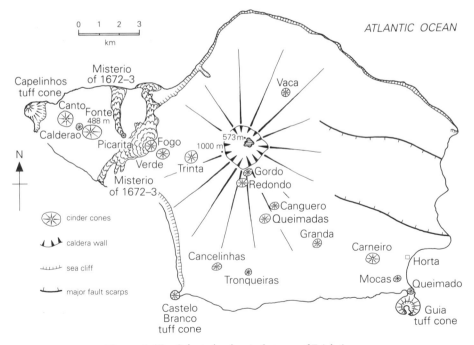

Figure 6.17 Selected volcanic features of Faial, Azores.

Figure 6.18 Capelinhos, with the Strombolian cone in centre, surrounded by the Surt-seyan tuff cone, much eroded by the sea on the left; wind-blown ash in the foreground.

perienced typical Surtseyan activity: a trailer to the birth of Surtsey in 1963. In a way, Capelinhos has been upstaged by Surtsey and its birthright has been stolen, because it had a prior claim to designating this type of eruption.

Capelinhos provides a superb example of the interplay between volcanic eruptions and shallow water, not only in the changing nature of the eruption itself, but also the interaction between the erupted materials and marine erosion. The typically Surtseyan initial eruptions of Capelinhos expelled weak tuffs that were quickly eroded; the subsequent Strombolian eruptions produced more consolidated fragments composed largely of spatter, and especially an armour-plating of lava-flows, that retarded marine erosion.

The eruption of Capelinhos was the longest since the Azores were settled in the 15th century, and lasted from 27 September 1957 until 24 October 1958 (Fig. 6.19). It had four distinct episodes. In the first two episodes, the eruptions were predominantly Surtseyan in character. In the third episode, Surtseyan activity was gradually superseded by increasingly dominant Strombolian activity. In the final episode, Strombolian activity occurred practically exclusively. Fumarole activity continued with decreasing temperatures and frequency until 1979.

Capelinhos first indicated its impending arrival with more than 200 weak earthquakes between 16 and 27 September 1957 that increased in incidence as their epicentres migrated to the westernmost point of Faial. At 08.00 on 27 September the sea began to boil in shallow water about 80 m deep, almost 1 km west of Capelinhos lighthouse and about 400 m west of the Ilheus dos Capelinhos. Gas and steam were emitted and the sea was discoloured during the next two days. The eruption began in earnest on 29 September with vigorous explosions of black basaltic ash in pointed jets, as if they had been fired from a gun. They were shaped like cypress fronds or

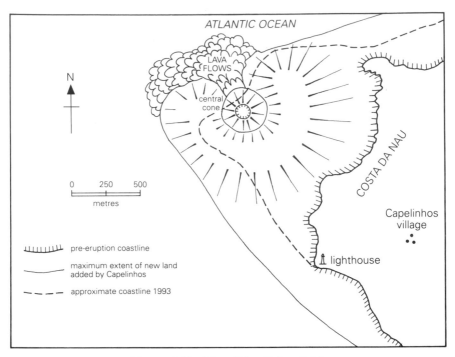

Figure 6.19 Map of Capelinhos, Faial.

cockerels' tails, which subsequently became recognized as the hallmark of Surtseyan activity. The jets rose 1 km into the air and were surrounded by billowing white clouds of steam that commonly rose 4 km high. The eruptions gradually focused on the western end of a fissure about 250 m long. The explosive spasms each lasted about half an hour and happened at two-hourly intervals. By 10 October 1957 a tuff cone, 100 m high and 800 m wide, had been built, almost enclosing a broad crater that remained open to the sea on the west, so that water had no difficulty in entering the vent. This activity ceased on 29 October 1957 and the weak, unconsolidated tuffs began to slump down the tuff cone, like a child's sandcastle swamped by the oncoming tide. By 1 November 1957, the tuff cone had been obliterated above sea level and redistributed in a broad shoal off western Faial.

The second episode began on 7 November 1957 after a week of total calm. It replicated the eruptive style of its predecessor, except that the vent was 500 m farther east. A new tuff cone grew around a wide crater open to the sea, and wind-blown ash built an isthmus joining the new volcano to Faial on 12 November 1957. This tuff cone was 600 m wide by mid-December, but it also began to slump into the sea as activity declined. It seemed as if the second tuff cone of Capelinhos would suffer the same fate as the first, until a red glow on its eastern flanks at 22.30 on 16 December 1957 revealed a change in eruptive style with the first lava emissions.

The third episode saw the advent of Strombolian eruptions. Lava-fountains and a lava-flow signified that sea water had failed to penetrate the vent for the first time, because of the copious accumulations around it. The third episode was a period of mixed activity. Surtseyan explosions, whenever water succeeded in entering the vent, were both more numerous and more voluminous. From time to time, notably on 29 December 1957, sectors of the tuff cone, composed largely of wet fragments, collapsed into the sea. Lava effusions were relatively infrequent, taking place especially on 31 January, 1 and 6 February, 6 March, and 20 and 23 April 1958, but the aa flows that formed an apron around the tuff cone were vital for the survival of Capelinhos because they retarded marine erosion and helped keep the sea from the vent. Throughout the winter, however, the Surtseyan explosions in-creased in violence, and by the end of March the tuff cone had reached 150m high and 1 km across, with a crater 450m wide, covering the Ilheus dos Capelinhos in the process. At this time, ash and bombs were often being fired 1800m into the air and they whistled down over the western promontory of Faial. The area became uninhabitable, especially when ash, mixed with sodium chloride evaporated from the ocean, destroyed crops and buried houses, and eventually formed a thick noxious blanket from which only the badly damaged Capelinhos lighthouse still emerged (Fig. 6.20).

The fourth episode was inaugurated by a phase, lasting from 12 to 14 May 1958, during which more than 450 tremors were recorded. They were not centred on Capelinhos, but on the crest of the Caldeira of Faial and at Praia do Norte on its lower northwestern flanks. Capelinhos thus did not cause these tremors. The western peninsula was warped up and down by more than 1.5m, fumarole activity was resumed in the caldera, and the earthquakes badly damaged Praia do Norte. The

Figure 6.20 Cabeços and the Costa da Nau, and the old Capelinhos lighthouse on the former coastal cliff, Faial.

effect on Capelinhos was radical. The Surtseyan eruptions ceased almost completely, because, except on rare occasions, the sea could no longer enter the vent. A large cone, composed predominantly of welded basaltic spatter, grew up from huge lava-fountains within the tuff cone. In addition, lava-flows issued from small external fissures, as well as from the central craters, accumulated in the atrium between the spatter cone and the tuff cone, and occasionally broke out to form an apron around the volcano (Fig. 6.19). This effusive activity ended abruptly on 24 October 1958. By then the core of Capelinhos was a steep-sided spatter cone, 160 m high, composed of wine-red basalt encircling a vertically walled, jagged crater. Around it spread the tuff cone, 150 m high and 1 km across, with gentler inner and outer slopes rarely exceeding 20. It was composed of innumerable layers of fine tuffs generally resembling buff sand and its surface was strewn with remarkable basalt bombs the size and shape of tortoises. These bombs were not quite solid on landing, so that their bases flattened on impact, whereupon molten lava burst from the solid carapace to form the head and legs of the tortoise. The seaboard flanks of the tuff cone were almost completely girded by lava-flows, and some 2.4 km² had been added to Faial.

Since the eruption, the life of Capelinhos has been dominated by its destruction by marine erosion. In spite of the efforts of its final effusive phase, Capelinhos did not emit enough lava to protect itself adequately and it has been eroded extremely rapidly. After 20 years it had already lost more than half its volume and only 1 km² now remains, and more than half the tuff cone and almost half the spatter cone within it have been removed. The jagged crater of the spatter cone now lies on the brink of vertical marine cliffs 150 m high. In the south, the sea has swept the tuff cone away to re-expose the chief islet of the Ilheus dos Capelinhos again in the cliffs. On the southern slopes of the tuff cone, the wind winnows the buff ash from the basaltic tortoise-shaped bombs and sets each one in relief. Even the site of the vent used during the first eruptive episode lies once again beneath the Atlantic Ocean. It is doubtful if much of Capelinhos will remain in a hundred years, unless another major eruption occurs to save it. However, if the trends of historic time are continued, it is by no means certain that this eruption will come in time.

Heimaey, 22 January 1973

The Vestmann Islands rise above a submerged plinth of erupted materials that stretches southwestwards from the zone of contemporary volcanic activity in Iceland (Fig. 5.5). Heimaey, the largest and oldest, covers only 16 km², and Surtsey, formed between 1963 and 1967, is now the second largest. All except the northern ridge of Heimaey, formed in the last ice age, were built above sea level during the past 10 000 years. However, apart from a small submarine eruption in 1896, the archipelago had not recorded volcanic activity since the settlement of Iceland in AD 874 until Surtsey erupted. Heimaey is dominated by the northern ridge which shelters Vestmannaeyjar, the main settlement on the island. The rest of Heimaey forms

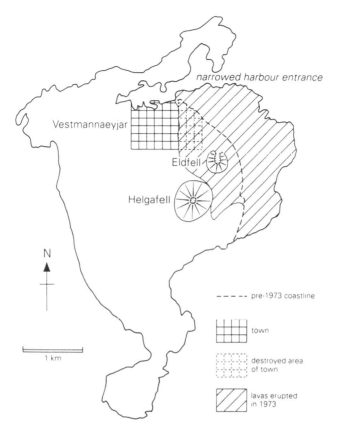

Figure 6.21 Map of Heimaey and the eruption of 1973.

a low plain rising southwards to two younger cinder cones, Saefell and Helgafell, which erupted just over 5000 years ago (Fig. 6.21).

When Surtsey erupted, the people of Heimaey were worried that volcanic activity would spread along the fissures to their own island. They had to wait just less than a decade. At 01.55 on 23 January 1973, lava-fountains forming a brilliant vermilion curtain burst more than 200m into the air along an 1800m fissure on the eastern slopes of Helgafell. The eruption was preceded by some 30 hours of earth-tremors repeated with increasing intensity and frequency, which gave warning that magma was rising to the surface. On 24 January, the lava-fountains waned and a Strombolian eruption began which expelled ash, lapilli and bombs, while basalts flowed eastwards to form a delta in the sea. Next day, activity had focused on one vent at the northern end of the fissure, where ash and lapilli, ejected every 1–3 seconds, rapidly formed a cinder cone and also showered down on Vestmannaeyjar, whose outskirts were less than 1km from the new volcano, Eldfell ("fire mountain"; Fig. 6.22). On 26 January the cone was already 100m high, and hot ash set fire to some houses and crushed others under its weight in the eastern parts of the town. The

Figure 6.22   Eldfell cinder cone, Heimaey (courtesy Margot Watt).

fissure also lengthened to 3km with renewed lava-fountaining, 150m high, that enabled lava to flow in a more threatening direction towards the town. At first, both flows and fragments were discharged at high rates and were often accompanied by carbon dioxide. By 30 January 1973, the cone already reached 185m, lava-flows had added 1km^2 to the east of the island, and 2 million m^3 of ash had fallen on Vestmannaeyjar. On 19 February, the northern sector of the crater wall of Eldfell collapsed and the breach directed yet more lava towards the town. From late February 1973, however, the eruption waned. Initial lava discharge rates of 100m^3 per second fell to 10m^3 per second in March, and emissions became so feeble by April 1973 that the end of the eruption was hardly noticed. But the lava-flows were still proceeding through the town to the harbour, whose entrance was now only 30m wide. It was only saved by human intervention and the end of the eruption in spring 1973.

In a sense, Heimaey was lucky. If the fissure had opened 1km farther north, for instance, Surtseyan explosions would have occurred in the harbour entrance and probably obliterated the harbour, the fishing fleet, and the town as well. Evacuation would then have been very difficult, and casualties probably inevitable.

The Strombolian eruption that occurred (with a moderate spread of fragments and rather slowly moving lavas) was of the kind where human initiative was likely to be most successful. Nevertheless, the disaster was well handled, because those in charge knew what to do. Organization and forethought left little place for panic. The whole population of 5300 on the island was evacuated to temporary accommodation on the mainland within seven hours, using the 77 fishing boats sheltering in the harbour from a storm that luckily had just ceased. The sick and the old were moved by air. Then the cattle and the cars, US$1 million worth of deep-frozen fish stored near the harbour, the money in the bank and municipal documents were transferred mainly to Reykjavík. A seven-person committee was set up to direct operations. The national bank released immediate funds for the evacuees, parliament increased taxes to pay for the expenses of the disaster, the government gave US$2 million, and more funds came in later from foreign governments and charitable institutions. From the onset of the eruption, only police and those with useful technical ability remained on the island, with suitable vehicles and equipment, although others were later allowed to return in organized groups to remove possessions from their threatened houses.

As the eruption developed, the dangers to be combatted were three-fold: poisonous gas, falling fragments, and lava-flows, which eventually assumed the greatest importance. Invisible, lethal, ground-hugging carbon dioxide swirled from the fissure and lurked in the cellars of Vestmannaeyjar for many days, often obliging those combatting the eruption to wear gas masks and avoid low-lying places.

The danger from falling fragments was more apparent in the first weeks of the eruption, when bombs and cinders damaged rooftops and broke windows, and hot ash often set fire to the houses and buried many others. The first palliative action undertaken, boarding up windows with metal sheets and shovelling ash from rooftop, could not keep pace with a prolonged eruption, although volunteers continued their efforts well into March. In the end, ash and lapilli burned and buried over 80 houses – and three-quarters of them were lost in the first week.

The first lavas flowed eastwards into the sea. If they had continued to flow in that direction, they would have created a few problems apart from warming the fish. But when the fissure lengthened on 26 January 1973 and part of the crater wall collapsed on 19 February, the lavas advanced on the eastern part of Vestmannaeyjar, threatening its harbour entrance, as well as the electric cables and the freshwater conduit from the mainland. The threat to the harbour was all the more important because Vestmannaeyjar was one of Iceland's major fishing ports, accounting for over one-eighth of its fish exports. Initial plans to bomb and breach the eastern flank of the crater, however, were abandoned because, if the west flank had also been disturbed, lava would have been directed straight to the town.

Early in February 1973, volunteers decided to pour sea-water onto the flow and consolidate its snout to slow down its advance. When this proved successful, the government sent out a ship in early March, whose pumps could dowse the lavas much more effectively. Powerful water pumps were then also dispatched from the United States and installed on land. Eventually, $20\,000\,m^3$ of lava was being congealed every hour and the flow stopped advancing. At the same time, ash was swept from the streets and compacted into barriers circumscribing the flow so that it could not spread westwards into Vestmannaeyjar. The preventive measures undertaken thus had some success. In fact, the lavas that reached into the harbour entrance and reduced its width to less than 30 m now provide better protection for the fleet.

But it is important to put the success into perspective. The methods used would only be successful with a plentiful water supply on gentle slopes threatened by slowly moving lavas, and without danger from violent explosions. The costs of pumping sea-water and building cinder barriers were high, both in terms of finance and human effort. But the people were determined to fight the volcanic onslaught on their lives. Nevertheless, in spite of all this endeavour, the electricity supply cable was cut, three of the five fish factories at the harbour were destroyed and the lava-flow alone claimed over 300 houses. The lava-flow was only halted, too, when the eruption was ending, when the flow had completed most of its advance, and when rates of discharge had already declined. But the lessons learned at Heimaey served as a basis for more difficult battles in the future. The damaged parts of the town have now been rebuilt and the population has almost reached its former total. Seismometers to register any rise of magma have been installed on the mainland and continuously recording tiltmeters have been set up on the island. Regular earthquake activity between Surtsey and Heimaey at a depth of about 20km, however, betrays the presence of magma that threatens more eruptions in the little archipelago.

Pinatubo, 14 June 1991

Pinatubo lies 80km northwest of the Philippine capital Manila in the volcanic chain of the Zambales Range (Fig. 6.23). Before the eruption, it rose to a height of 1745m and it had apparently shown no signs of life since the Philippines were first settled by the Spaniards in 1541, although it had been mapped as an active volcano in 1988. It is a measure of Pinatubo's apparent harmlessness that the Clark American Air Force Base lay within sight of the volcano, and the naval base of Subic Bay was only 40km away to the southeast. Pinatubo gained new respect in 1991 by unleashing the most violent eruption for perhaps a century (Fig. 6.24). As some 15000 people inhabited its fertile slopes with about half a million more on its perimeter, it was almost certainly only through a large-scale evacuation that Pinatubo was prevented from rivalling the death-tolls of the most lethal eruptions in the same period.

The summit of Pinatubo was formed by a large dacitic dome, 3km across, and its flanks were coated with old ash and pumice. Although little else was known about Pinatubo, the unexpected eruption, so near a strategic base, stimulated rapid geological investigations. The Philippine Institute of Volcanology and Seismology established that activity had started on Pinatubo about 1.1 million years ago and that the last three eruptions had taken place about 400–600 years ago, between 2500 and 3000 years ago, and between 4400 and 5100 years ago. Sections exposed along deep gullies revealed that the ash and pumice, extending over 20km from the summit, had been deposited chiefly by nuées ardentes and lahars. The arousal of the volcano alerted experts and authorities alike to the possibility that such potentially lethal eruptions might be repeated. Fortunately, the appropriate palliative measures were undertaken in co-operation between the Institute of Volcanology, the Philippine civil authorities and the United States military authorities, so that the story of the eruption of Pinatubo is dominated by successful disaster management, just as events at Nevado del Ruiz had highlighted its failure.

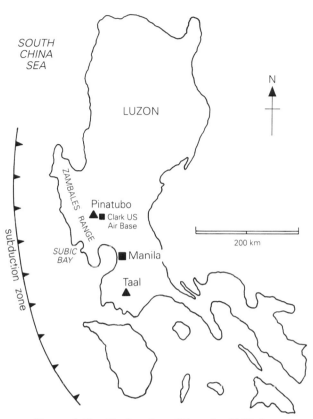

Figure 6.23 The location of Pinatubo, Phillipines.

The first eruption of Pinatubo in historic time began in the afternoon 2 April 1991 from a fissure, 1.5 km long, that transected the northern flank of the volcano and its summit dome. It was a phreatic eruption that formed a small crater row, and coated villages with ash 10 km away to the southwest. On 3 April, the fissure lengthened and that same afternoon 5000 people were evacuated from the most dangerous zones situated within 10 km of the summit of Pinatubo. Seismographs were installed on the mountain and were linked telemetrically to computers at Clark Air Base, where a volcanic observatory was established.

In April and early May 1991, geological surveys revealed the violent history of Pinatubo. A volcano hazard map of the area, prepared by the Philippine Institute of Volcanology, was published on 23 May 1991, and it brought out the probable dangers from lahars and nuées ardentes (Fig. 6.25). It corresponded well with the actual course of events the following month. A video-film of volcanic hazards prepared by the French geologists Maurice and Katia Krafft was not only shown to the authorities, but also appeared several times on television. Ordinary people as well as the civil and military powers thus soon became fully aware of the enormous threat that Pinatubo could represent. (It is a tragic irony that the Kraffts were themselves

Figure 6.24 Pinatubo in eruption.

killed by an unexpected nuée ardente on Unzen in Japan on 3 June 1991.)

Throughout May 1991, the seismometers registered some 1800 small earthquakes as the magma rose from its shallow reservoir, none of which exceeded a magnitude of 2.5 on the Richter scale. They were concentrated between 2 km and 6 km deep about 5 km northwest of the summit. Measurements at the new vents also showed that sulphur dioxide emissions from the rising magma increased ten-fold between 13 May and 28 May; but these emissions declined again in the first two weeks of June, apparently because the rising magma had itself sealed the routes by which its gases escaped to the surface. From 1 June 1991, the volcanic earthquakes changed to a new focus, which was less than 5 km deep and much closer to the fumarole vents, 1 km northwest of the summit. The change suggested that magma was now breaking open a conduit through the solid rock. On 3 June a small explosion heralded a whole series of ash eruptions, increased earthquakes and the start of episodes of harmonic tremor caused by more sustained magma ascent. As a result, on 5 June, surveillance teams forecast a major eruption within two weeks. On 6 June 1991, the tiltmeter near the summit revealed that the volcano was bulging outwards at an increasing rate, as more earthquakes occurred. These manifestations were brought to a halt when a column of ash and steam rose almost 8 km in height

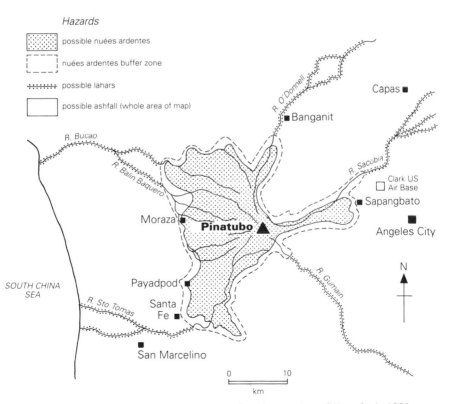

Figure 6.25 The hazard map prepared before the eruption of Pinatubo in 1991.

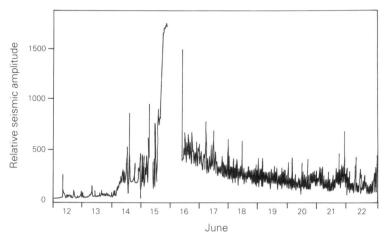

Figure 6.26   The relative seismic amplitude of explosions at Pinatubo, 12–22 June 1991; there were no recordings during the climax of the eruption.

on 7 June. The surveillance team at once announced that a larger eruption was possible within 24 hours and thus recommended further evacuation from the flanks of the volcano.

Magma reached the surface just northwest of the summit on the morning of 8 June 1991, not in an explosion, but as a dome, less than 150 m across. This dome emitted small nuées ardentes from 9 June onwards. Minor ash emissions accompanying harmonic earth-tremors indicated that magma continued to rise until 12 June. Andesitic cinders were ejected along with the predominantly dacitic ash, which indicated that the eruption was precipitated when the new hot basalts invaded the reservoir containing cooling dacite left over from the last eruptions more than 400 years ago. By 9 June 1991, when a major eruption was obviously imminent, the authorities extended the evacuation zone to include all those living within 20 km of the summit of the volcano, thus increasing the number of local evacuees to over 25 000. Next day, 10 June, 14 000 Americans were also evacuated from Clark Air Base, along with their aircraft, leaving behind three helicopters and 1500 maintenance and security personnel.

On Philippine Independence Day, June 12, Pinatubo entered a more powerful phase. Major eruptions occurred at 08.51 on 12 June, 22.52 on 12 June, and 08.41 on 13 June. Each was heralded by swarms of sustained earthquakes lasting a total of up to four hours (Fig. 6.26). The eruption at 08.51 on 12 June, for instance, lasted 35 minutes and sent a Plinian column 19 km into the air at a speed exceeding 60 km an hour. Ash winnowed by the winds from the column forced Manila airport to close, while several small nuées ardentes invaded already-evacuated villages in the valleys radiating from the summit. The evacuation zone was extended to a radius of 30 km from the volcano, increasing the total number of local evacuees to more than 58 000; and 600 more people were removed from Clark Air Base.

Observation of Pinatubo then became increasingly difficult, as the ash clouds grew more widespread and the weather deteriorated with the arrival of Typhoon Yunya. There was a lull in the proceedings between the eruption at 08.41 on 13 June until 13.09 on 14 June, when the climactic phase began in earnest.

From a number of vents operating simultaneously, short but violent eruptions, generating narrow Plinian columns, continued during the afternoon and evening of 14 June, at 13.09, 14.10, 18.53 and 23.20, and at 01.14, 02.57, 05.55, and 08.09 on 15 June. The columns usually exceeded 20km, and perhaps reached 40km in height. But the explosion at 05.55 was predominantly lateral and seemed to indicate that the summit zone was about to collapse. Many nuées ardentes, extending as much as 15km from their source, were emitted at this time. The remaining occupants of Clark Air Base were therefore evacuated to a safe haven.

Eruptions of increased intensity, often accompanied by vigorous nuées ardentes, were registered on the remaining seismograph in operation at 10.27, 11.17, 11.55, 12.21, 12.52 and 13.42 on 15 June (Fig. 6.26). The volcano, in fact, almost experienced a single continuous eruptive convulsion during the late morning of 15 June 1991. On that afternoon, the seismographic equipment was destroyed and visibility all around Pinatubo was reduced almost to zero as Typhoon Yunya brought gales and heavy rains to the whole area. A series of strong, widespread earthquakes, ranging in magnitude from 4.8 to 5.6 on the Richter scale began at 15.39 on 15 June and continued throughout the following night. They were caused when vast quantities of ash and pumice were expelled and a caldera, 2km across, collapsed 1km north of the former summit. The volcano lost over 300m in height. Spread unusually widely by Typhoon Yunya, the ash formed a mantle more than 5cm thick, covering about 4000km^2 around Pinatubo (Fig. 6.27). The deposits of the climax were dacitic, beginning with layers of fine grey ash, followed by coarse pumice, and then by fine, sandy pumice and ash. Crops and vegetation were buried, and the ash, made abnormally heavy by the typhoon rains, caused many earthquake-weakened rooftops to collapse, thereby provoking most of the casualties during the eruption. Without the intervention of Typhoon Yunya, the death-toll from the eruption would certainly have been much smaller, because of the efficient evacuations in the days preceding the climax.

After the summit caldera collapsed, many nuées ardentes were expelled down the radial valleys and extended as far as 16km from the vents. Their dacitic deposits averaged 30–50m in thickness, and even reached 220m in places. Their magma volume has been estimated at between 5km^3 and 7km^3. The climactic phase of the eruption waned rapidly during the evening of 15 June 1991. However, ash was exploded regularly, in columns often reaching 10km high, until early September 1991, while a dome 300m across and 100m high grew up in the fuming caldera. As the major eruption declined, the rains from Typhoon Yunya continued to drench the unconsolidated newly expelled ash and pumice, and formed lahars that rushed down almost every major valley radiating from Pinatubo, causing over a dozen deaths, burying valuable crops, and severely damaging houses and bridges alongside the River Abacan in Angeles City. Almost as soon as the typhoon had abated, the monsoonal rains began in early July, which generated further destructive lahars.

119

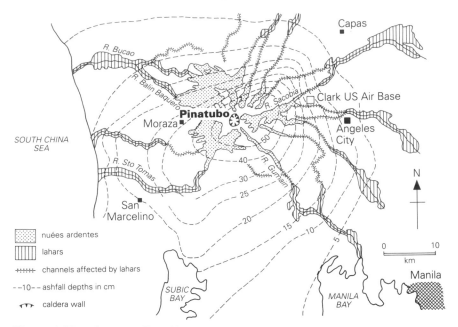

Figure 6.27 The areas affected by nuées ardentes, lahars and ash from Pinatubo in June 1991 (cf. Fig. 6.25).

These secondary lahars (caused by rain) are now closely monitored, but will continue to be a threat during every monsoon until the vegetation resumes its firm foothold on the slopes of Pinatubo in future years. Some 400 villages are still in danger.

In all, some 86 000 ha of agricultural land were affected by ashfalls and lahars, and about 650 000 people were forced out of work for several weeks at least; and 94 per cent of the Aeta tribe were displaced from their traditional homelands on the slopes of Pinatubo. In spite of the success of the timely evacuation of the danger zone, about 300 lives were lost during the eruption, and over 500 people died from diseases in the evacuation centres.

The eruption had two other, more widespread effects. Pinatubo ejected an estimated 7–14 km^3 of ash, dust and aerosols into the air. The ash clouds presented an aviation hazard because they were indistinguishable from normal high-altitude clouds. Nine airliners were forced to make emergency landings when the ash damaged their engines. Pinatubo also expelled unusually high volumes of sulphur dioxide into the stratosphere. These eventually formed sulphuric acid droplets which reflected solar radiation, and therefore lowered the amount of the Sun's heat able to reach the Earth's surface.

A month after the eruption of 15 June 1991, the Total Ozone Mapping Spectrometer on the Nimbus-7 American meteorological space satellite revealed that a dust and sulphur dioxide aerosol cloud had encircled about 40 per cent of the Earth's surface between latitudes 25° north and 20° south, and it had encompassed the whole globe before the end of 1991. Pinatubo had expelled between 10 and 20 million tonnes of sulphur dioxide into the stratosphere, double the volume ejected by El Chichón in 1982, and about one-quarter of the sulphur dioxide annually discharged by the world's industries. It formed probably the densest veil seen in the Northern Hemisphere since Krakatau erupted in 1883. Preliminary studies suggest that average annual global temperatures could be reduced by as much as 1.5°C, at least for part of the period from 1991 to 1995.

It is possible that the first climatic effects of the eruption of Pinatubo were felt in the cool, wet Canadian summer of 1992, when unprecedented amounts of pack-ice remained unmelted in Hudson Bay; in abnormally early snowfalls in September and October 1992 in Alaska, Scotland, Switzerland and Moscow; in the coldest October experienced for years in Britain in 1992; and in the floods associated with the stormy summer and autumn in 1993 over much of Europe and North America. Although aerosols and dust veils are only two of many ingredients in the cocktail of short-term climatic oscillations, it is likely that Pinatubo will be held responsible, perhaps quite correctly, for these unpleasant events. It is, however, still too early to assess the overall results of the eruption, for its geological, geographical, climatic and social dimensions still await the results and perspectives of future research. It is already clear, however, that this mighty outburst was only prevented from being extremely lethal by timely and efficient evacuation of the threatened population living on its flanks.

Lanzarote, 1730–36

Lanzarote is the northeasternmost of the Canary Islands and it covers an area of about 795 km². Its higher parts form plateaux of basaltic lavas, between which stretches the central saddle of the island, often lying below 300m and pock-marked by lines of cinder cones from which issue a multitude of lava-flows. The older volcanic areas are covered with a weathered, pale ochre crust, the "caliche"; much of central Lanzarote is blanketed with layers of black ash and lapilli called "picon", in which farmers have excavated hollows and planted crops; and cones and lava-flows erupted between 1730 and 1736 form a harsh area covering about 200 km² in the northwest (Fig. 6.28). Practically all the eruptions of Lanzarote sprang from fissures, which built up the island from a depth of 2500m on the Atlantic Ocean floor. Although the first activity above sea level in Lanzarote occurred about 11 million years ago, 80 per cent of the island is less than half a million years old, and, apart from a minute outcrop of trachytic rocks, all the lavas are basaltic.

It is the 18th-century eruptions that have added the spectacle to the landscape of Lanzarote. Between 1 September 1730 and 16 April 1736, the island was the scene of one of the greatest European fissure eruptions in historic time, in terms of its output volume and duration. No previous historic eruptions had taken place on the island and the subsequent calm has since been broken only by three very brief outbursts in 1824. One-quarter of Lanzarote was given an entirely new landscape and farms and villages were buried completely in the west-central parts of the island (Fig. 6.29). The black, grey or reddish-brown lavas in deep craters, dark pits, jagged

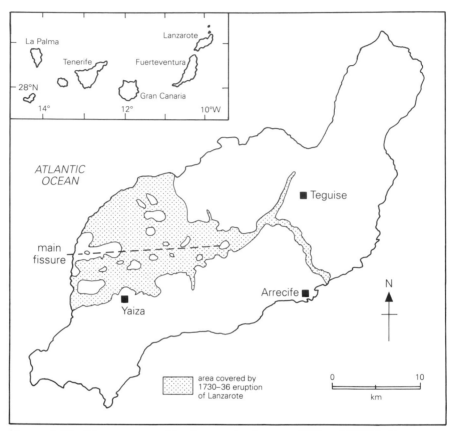

Figure 6.28 The area covered by the eruption between 1730 and 1736 in Lanzarote, Canary Islands.

ridges, smooth ash cones or rugged flows give a remarkably brutal impact to this, the starkest landscape in the Canary Islands, which has scarcely a twig of vegetation growing upon it (Fig. 6.30).

Activity took place along a belt of fissures some 4km wide, stretching over a total distance of 18km. It occurred in five distinct phases and formed over 30 cinder cones, many spatter cones and hornitos, as well as widespread basaltic flows (Figs 6.31, 6.32).

The first phase had two separate parts. Activity lasting from 1 September until 19 September 1730 constructed the Los Cuervos cone, 8km northeast of Yaiza, and emitted flows that reached the northwest coast. After three weeks of tranquillity, the outburst resumed on 10 October, some 1.5km to the north. Fluid lavas poured forth again, covering most of those emitted from Los Cuervos, but also spreading towards Yaiza. At the same time, the Santa Catalina cone erupted over the destroyed village and expelled the largest volume of fragments of the whole eruptive period. A group of vents 1km to the northwest destroyed Mazo. They formed Pico Partido ("clo-

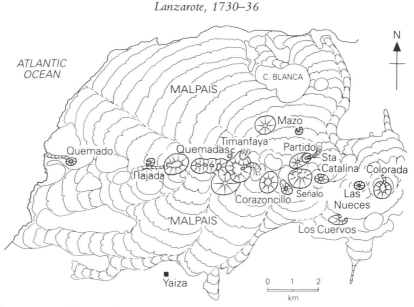

Figure 6.29 The main formations of the eruption between 1730 and 1736 in Lanzarote.

Figure 6.30 Pico Partido and an incompletely formed lava tube, Lanzarote.

ven"), a cone almost 200m high, whose jagged cloven black outline is the most striking in Lanzarote (Fig. 6.30). It is crowned by a large spatter cone, with three aligned craters, and flanked by a tuff ring, bordered by the remnants of lava cascades, and containing a congealed lava lake, itself drained by an incompletely roofed tube,

123

Figure 6.31   Fissure eruptions in 1731 in Lanzarote; looking from the crater of Montaña Rajada to the Montañas Quemadas with Timanfaya beyond.

emphasized by lichen growth, that winds its way down the cone's lapilli-strewn flanks. Like the olivine xenoliths scattered on its slopes, Pico Partido is a collector's item for admirers of basaltic forms (Figs 6.33, 6.34).

The second phase may have begun on 7 March 1731 – and certainly by 20 March and it lasted until June 1731. Activity concentrated on four vents that formed the Montañas del Señalo, less than 1 km south of Pico Partido (Fig. 6.35). It is possible that Corazoncillo, a distinctive cone of rose-pink lapilli 150m high, was also formed at this time. The more viscous lavas of this phase rarely exceeded 5 km in length, but still managed to approach the outskirts of Yaiza.

The third phase lasted for six months. Activity suddenly switched 12 km westwards along what was to become the main fissure. It began in June 1731 with an offshore Surtseyan eruption that terrified the population but failed to build a permanent islet. Soon afterwards eruptions began on land and progressed intermittently eastwards along the fissure. Successive vents formed the cinder cones of El Quemado ("burnt") in June, Montaña Rajada ("split") in July and the four Quemadas cones between October and December 1731 and January 1732, and their lavas now coat much of western Lanzarote. It was these lava-flows that caused the people of Yaiza to abandon their homes for Gran Canaria early in January 1732.

Activity of the fourth phase was confined to the middle section of the main fissure at Timanfaya from the winter of 1732. As no documentary evidence has been brought to light, it is uncertain how long this phase lasted, but it seems to have continued for at least a year – and probably simmered for much longer. It constructed

124

Figure 6.32 Hornito: the Virgin's Mantle, Lanzarote.

the large intersecting reddish cinder cones of Timanfaya, many spatter cones, and lava-flows that spread both northwards and southwards.

The fifth phase of the eruption saw a farther eastward shift along the main fissure in late March 1736. The small Las Nueces cinder cone was formed in a few days and very fluid lavas gushed out which made their way to the southeast coast 20 km away. In early April, eruptions lasting ten days built the Colorada ("red-coloured") cinder cone 1 km farther east, and sent out flows that almost reached the north coast. This eruption brought 5½ years of activity in Lanzarote to a close on 16 April 1736.

The early stages of the eruption were recorded in two independent sources. The first, the diary of the parish priest of Yaiza, is now lost, but survives, summarized, in translation in German and French. The second source is the correspondence, preserved in the Spanish national archives, between the Royal Court of Justice of the Canary Islands and the committee established on Lanzarote to manage the crisis. The diary ended on 28 December 1731 and the archive correspondence on 4 April

125

1731, so that only the initial stages of the eruption were recorded directly.

For Father Andres-Lorenzo Curbelo, the parish priest of Yaiza, it was like this:

On the first of September 1730 between nine and ten in the evening, the earth suddenly opened up near the village of Timanfaya, two leagues [in fact, 8km] from Yaiza. During the first night an enormous mountain [Los Cuervos] rose up from the bosom of the Earth and it gave out flames from its summit for 19 days. A few days later, a fissure opened up probably at the foot of the newly formed cones, and a lava-flow quickly reached the villages of Timanfaya, Rodeo and part of Mancha Blanca. This first eruption took place east of the Montaña del Fuego, half-way between that mountain and Sobaco. The lava flowed northwards over the villages, at first as fast as running water, then it slowed down until it was flowing no faster than honey. A large rock arose from the bosom of the Earth on 7 September with a noise like thunder and it diverted the lava flow from the north towards the northwest. In a trice, the great volume of lava destroyed the villages of Maretas and Santa Catalina lying in the valley. On 11 September, the eruption began again with renewed violence. The lavas began to flow again, setting Mazo on fire and overwhelmed it before continuing on its way to the sea. There, large quantities of dead fish soon floated to the surface of the sea, or came to die on the shore. The lavas kept flowing for six days altogether, forming huge cataracts and

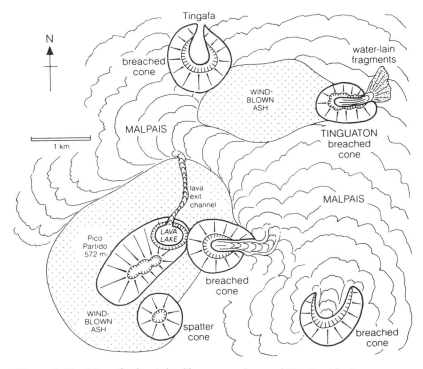

Figure 6.33   Map of volcanic landforms on and around Pico Partido, Lanzarote.

126

Figure 6.34 Congealed lava lake within a tuff ring formed on the northern flanks of Pico Partido in 1730–31.

Figure 6.35 Spatter forming the crater edge of Pico Partido, looking south to the Montañas del Señalo.

making a terrifying din. Then everything calmed down for a while as if the eruption had stopped altogether. But, on 18 [in fact, 10] October, three new openings formed just above Santa Catalina, which was still burning, and gave off great quantities of sand and cinders that spread all around, as well as thick masses of smoke that belched forth from these orifices [Santa Catalina and Pico Partido] and covered the whole island. More than once, the people of Yaiza and neighbouring villages were obliged to flee for a while from the ash and cinders and the drops of water that rained down, and the thunder and explosions that the eruptions provoked, as well as the darkness produced by the volumes of ash and smoke that enveloped the island . . . On 28 October, the livestock all over the nearby area suddenly dropped dead, suffocated by an emission of noxious gases that had condensed and rained down in fine droplets over the whole district. Calm returned on 30 October.

Ash and smoke started to be seen again on 1 November 1730, and they were erupted continually until 10 November, when a new lava-flow appeared, but it only covered those areas that had already been buried by previous flows. On 27 November, another lava-flow [from Pico Partido] rushed down to the coast at an incredible speed, reached the shore on 1 December and formed a small islet that was soon surrounded by masses of dead fish. On 16 December the lavas changed direction and reached Chupadero, which was soon transformed into what was no more than an enormous fire. These lavas then ravaged the fertile croplands of the Vega de Uga [1 km east of Yaiza]. On 7 January 1731, new eruptions [from Pico Partido] completely altered the features formed before. Incandescent flows and thick smoke emerged from two openings in the mountain. The clouds of smoke were often traversed by bright blue or red flashes of lightning, followed by thunder as if it were a storm. On 10 January 1731, we saw an immense mountain rise up, which then foundered with a fearsome racket into its own crater the self-same day, covering the island in ash and stones. Burning lava-flows descended like streams across the malpaís as far as the sea. This eruption ended on 27 January 1731. On 3 February, a new cone grew up [Montaña Rodeo] and burned the village of Rodeo. The lavas from this cone crossed the whole area and reached the sea. These lavas continued to flow until 28 February. On 7 March, still further cones were formed, and their lavas flowed north towards the sea and completely destroyed the village of Tingafa. New cones with craters [Montañas del Señalo] arose on 20 March and continued to erupt until 31 March. On 6 April, they started up again with even greater violence and ejected a glowing current that extended obliquely across a previously formed lava-field near Yaiza. On 13 April, two mountains [of the Montañas del Señalo] collapsed with a terrible noise. On 1 May, the eruption seemed to have ceased, but, on 2 May, a quarter of a league farther away, a new hill arose and a lava-flow threatened Yaiza. This activity ended on 6 May and, for the rest of the month, this immense eruption seemed to have stopped completely. But, on 4 June, three openings occurred at the same time, accompanied by violent earthquakes and flames that poured forth with a terrifying noise, and once again plunged the inhabitants of the island into great consternation. The orifices soon joined up into a single cone of great height from which exited a lava-flow that rushed down as far as the sea. On 18 June a new cone was built up between those that already masked the ruins of the villages of Mazo, Santa Catalina and Timanfaya. A crater opened up on the flanks of this new cone that started to flash and expel ash. The cone that had formed above the village of Mazo then gave off a white gas the like of which nobody had ever seen before. [The third phase began on the west coast.] . . . Then, about the end of June 1731, the whole west coast was covered by enormous quantities of dead fish of all kinds, including some that had never been seen before. These eruptions took place under the sea. A great mass of smoke and flames, which could be seen from Yaiza, burst out with violent detonations from many places in the sea off the whole west coast. In October and December, further eruptions [of the Montañas Quemadas] renewed the anguish of the people. On Christmas Day 1731 the whole island was affected by the most violent of all the earthquakes felt during the previous two [sic] years of disasters. On 28 December a new cone [in the Montañas Quemadas] was formed and a lava-flow was expelled from it southwards towards Jaretas. The village was burned and the Chapel of San Juan Bautista near Yaiza was destroyed.

Seeing Yaiza so closely threatened from yet another direction and despairing of ever seeing an end to their travail, the villagers decided to take refuge in Gran Canaria. But Yaiza, in fact, just remained on the edge of the new lava-flows.

The recently discovered correspondence in the Spanish national archives at Simancas offers an administrator's perspective on the eruption and the problems presented in its management (Archivo de Simancas (Spain), Gracia y Justicia, legajo 89). Power in the Canary Islands was in the hands of the Governor General, residing in Tenerife, and the Regente of the Royal Court of Justice (the Real Audencia), established in Gran Canaria. In Lanzarote, power devolved onto the Chief Administrator (or Alcalde Mayor). Soon after the eruption began, a committee (or junta) was set up in Lanzarote to help the Alcalde Mayor manage the crisis. As the eruption progressed, however, the Regente lost patience with the junta and appealed to the Crown, sending the correspondence (with a summary of events) to Madrid.

The summary letter of 4 April 1731 from the Regente to the Crown describes the start of the eruption:

On the first of September of last year, a volcano erupted on the island of Lanzarote that was prodigious both in the quantities of fire, stones [cinders and lapilli] and sand [ash] expelled, and in its duration – for it still continues to destroy the island today. The fire was so strong and so high that it could continually be seen, from this island [Gran Canaria] and others for a distance of more than 50 leagues [*c.* 200 km]. The stones were in such quantity and of such size that they formed many high mountains during the eruption, and a noise like thunder was heard whenever they entered the water. For several days the eruption shook buildings, doors and windows as far away as the mountains of this and other islands. The quantity of sand was such that, as well as causing as much, if not more damage than the fire to houses, lands and water-cisterns, it also formed an islet deep in the sea which was almost one league [*c.* 4km] long and over three leagues [*c.* 12km] in circumference. The sand was also blown by the wind and the fire as far as 15 leagues [*c.* 60km] away to the interior of Fuerteventura, without, however, causing notable damage there. The eruption was only interrupted for ten days in September before it started again with renewed vigour from two new mouths, but the original mouth only spat forth smoke and sulphur thereafter. The second interruption lasted seven days in early January. It was followed on 20 January [1731] by the opening of a fourth mouth, thereby intensifying the power, thunderous noises and damage caused. The other mouths then ceased. All the mouths opened within a distance of a little over a league and the fire [lava-flow] was always directed to one side.

From the start, the upper échêlons of the administration had taken the lead in managing the crisis created by the eruption. It is a recurrent theme of the dossier that the Regente was concerned both with public order and mitigating the effects of the disaster on the population, adapting policy and allowing more evacuation as the effects of the eruption were increasingly felt. The junta in Lanzarote, on the other hand, not only procrastinated in implementing the Regente's decrees, but became increasingly pre-occupied with the red herring of the military defence of the island.

Believing, no doubt, that the eruption would be as brief as those between 1704 and 1706 in Tenerife, the Regente's first ordinances were designed to prevent panic and protect the grain stocks both from the eruption and theft. Then, on 17 October 1730, the junta in Lanzarote informed the Regente that two new vents had opened about 17.00 on 10 October,

> . . . three-quarters of a league from the first volcano [Los Cuervos] and the distance of a good musket shot apart [*c.* 600m]; one was very close to the already burned church of Santa Catalina [Santa Catalina cone], the other near Mazo [Pico Partido] . . . So much fire and sand was expelled that the trouble was felt three or four leagues [12km or 16km] away, damaging roofs and fields . . . The Vega [fertile plain] of Tomaren, the heart of the island, and many vegas belonging to adjacent villages, as well as to individual owners – all the best land in the area – are henceforth lost because of the sands. . . . [and] in our misfortune . . . birds, rabbits, mice and other little animals run wild over the volcanic sands without any means of feeding themselves, although all that is but nought compared with the pain caused by the tears and laments of men, women and children under the onslaught of this hostile element.

To all this was coupled the continual shaking of the island which caused probably the greatest terror – "especially amongst the womenfolk". In late October, as Santa Catalina and Pico Partido continued their vigorous Strombolian outbursts, it became clear that the situation was worse than any experienced in the Canary

Islands in 300 years of Spanish rule. Faced with a possible collapse of public order and a famine, the Regente decreed on 31 October 1730 that no families be allowed to leave for other islands on pain of severe penalties, both for the people and the boat-owners transporting them; that under no pretext whatsoever should grain be removed without permission from the island; that a census be undertaken forthwith of the food, water resources and arable land on Lanzarote; that the grain stocks and herds whose protection could not be assured should be transferred to Fuerteventura; and that the poor and dispossessed should be given alms and shelter in Lanzarote. At the same time, the membership of the junta was increased. It comprised the Great and the Good of Lanzarote: "six deputies, two judges, a churchman (the Commissioner of the Holy Inquisition), a soldier and two men of honour and conscience". All but two, however, were military men, and reacted in military ways.

As the eruption continued with vigour throughout the autumn of 1730, and the junta increasingly focused its attention on defence, the Regente realized the plight of the population. A map, commissioned by the military governor of Fuerteventura (not Lanzarote!) was painted in oils and sent to the Regente on 18 November 1730 and perhaps brought home to him the extent of the damage in Lanzarote.

On 9 December 1730, the Regente therefore decreed that the poor and homeless should now be free to depart, but only for Fuerteventura, with their grain and chattels. Enough people should, however, remain on Lanzarote to ensure its defence against any invasion. (The Spaniards perhaps suspected that the British might be wily enough even to invade such a shaking, ash-covered island. In 1728, Spain had failed to recover Gibraltar after a 14-month siege.) The junta also reckoned on 29 December 1730 that, given the grain supplies recently assessed, "at very most 600 families could be maintained on Lanzarote and thus the other 400 families that could not be nourished must needs leave". The junta requested that they should be allowed to go to other islands, because limiting the evacuation solely to Fuerteventura would place too great a strain on that island and expose it to famine and ruin. In the same letter, the junta described how "The mouth near Mazo gave out fire that split into several branches, flowing as fast as the Betis [the River Guadalquivir], damaging everything in their paths and ending up in the sea . . . [and] piles of incandescent rocks were dragged along in the flames vomited by the Infernal Fire-Dragon that is destroying the island."

In response, on 29 January 1731, the Regente decreed that 400 families – with their goods and grain – be allowed to leave for other islands in addition to those granted permission to go to Fuerteventura.

Although a controlled evacuation of Lanzarote was becoming increasingly probable and acceptable, the junta, on 19 February 1731, described the devastation of the island, where successive volcanic orifices were exploding and blanketing more and more valuable land beneath the ash, so that it could no longer yield food for man nor beast. The junta nevertheless recommended that a garrison of 200–300 men should be left in Lanzarote, to prevent an enemy occupation.

But, as the Regente told the junta on 25 March 1731, "it was irrational to sacrifice produce and population to defend the island and thereby probably lose everything". His orders became increasingly blunt and peremptory; and in the same letter he declared: "that without protest, interpretation, or sundry other delay, boats should be brought so that all those families . . . together with their remaining grain, could leave for the island of their choice, as had been ordered already . . . [and] that the boat owners should comply, without fail, with the contracts established with the citizens for the evacuation of their persons, goods and grain."

On 4 April 1731, the Regente appealed directly to the Crown and sent the whole dossier to Madrid. The details of the sequel remain unknown. The eruption was only seven months old and had just begun its second phase when the dossier comes to an end. The diary of the priest at Yaiza, too, is less precise after the spring of 1731. It is clear, however, in these early and terrifying months that the people continually asked for succour, and for permission to depart with their property (and especially with the grain and herds that were their principal sources of wealth). The Regente reacted correctly to the increasing gravity of the emergency and came around to the same view, but the stumbling block was the junta in Lanzarote, who seemed to lack the imagination to cope with events.

When the eruption finally ended in 1736, a quarter of the island had been resurfaced with between $3\,km^3$ and $5\,km^3$ of lava-flows and fragments. Some of the best land on Lanzarote had been destroyed, many inhabitants had fled, and 26 settlements had been overwhelmed. Famine, hardship and great alarm were widespread, but no lives were lost directly as a result of the eruption.

In the aftermath, however, the people fought back by discovering new ways of farming. Although the lava-flows were intractable, the "picon" of fragments was soon exploited by remarkable agricultural techniques – crops were planted in small hollows to protect them from the desiccating Trade Winds from Africa; the picon itself absorbs and retains the modest ambient humidity; extensive underground channels draw groundwater to the surface; and crops have been diversified and specializations developed. Thus, the population of Lanzarote, which had stood at 4967 at the eve of the eruption had practically doubled to 9705 by 1768.

Now the area around Timanfaya, Quemadas and Montaña Rajada is a national park with restricted access. Elsewhere, the volcanic landscape can be visited freely on foot or on a dromedary, according to taste. The Islote de Hilario, near Timanfaya, still registers temperatures of 600°C about 13m below ground, and the fumaroles emitted are used for cooking in the adjacent restaurant.

PART III

Volcanic landforms

7

Lava-flows

The eruption of Etna in 1669

The largest and most famous eruption recorded on Etna during historic time began on 11 March 1669 and ended 122 days later on 11 July. From 25 February 1669, earthquakes that severely damaged Nicolosi heralded the opening of a fissure, 12 km long and 2 m wide, on 11 March that stretched like a gaping wound from just north of Nicolosi at 850 m to a height of 2800 m at the foot of the summit cone. During the night a vent started to build the twin Monti Rossi cones, and the fissure spewed forth hot, fluid basalts which overwhelmed the village of Belpasso. These lavas emerged at a rate of 96 m^3 per second, almost four times the average discharge of flows on Etna. Within a few days, village after village was threatened and swamped. S. Giovanni di Galermo (6 km from the Monti Rossi) was reached on 15 March, and Misterbianco (9 km away) on 25 March. As the Monti Rossi cinder cones quickly grew up on the fissure, the lava-flow divided into three arms, and the main one advanced towards Catania. With courage and scientific acumen, Diego Pappalardo and 50 Catanians diverted the flow by opening its solidified sidewalls so that the molten lava in the axial channel would turn sideways. Unfortunately, this new branch advanced towards Paterno, whose alarmed inhabitants drove off the Catanians before they could cause more trouble. This first known scientific attempt at volcanic hazard control was thus thwarted, because the new breach healed again and the flow resumed its original course. Nine villages had been destroyed by the time the lavas reached Catania, 12 km from the vent, on 12 April 1669. In AD 252 lavas had been arrested at the gates of Catania when the veil of the recently martyred St Agatha was waved at their advancing snout. Now the veil proved impotent. During the next three days the lavas wrapped around and rose above the city walls, pushing one stretch down on 15 April. The flow reached the sea south of Catania on 23 April and by 30 April had overwhelmed most of the western part of the city. No lives were lost in this eruption, because, after all, the lavas had travelled an average of only 500 m a day. But they had advanced inexorably, sustained by their rapid discharge. It is doubtful if Pappalardo and his men could, in fact, have saved Catania without diverting the whole flow onto Paterno. When the eruption ceased on 11 July 1669, the lavas had covered an area of 37.5 km^2, with an average thickness of 25 m. Some 937.5 million m^3 of lava had been emitted at an average rate of 89 m^3

per second. And the Monti Rossi had reached a height of 250m and still rank amongst the highest cinder cones on Etna. After this *tour de force*, the flanks of Etna saw little activity until 1763.

Lava effusions

Lava flows across the Earth's surface when fluid magma erupts and runs down slope under the influence of gravity. Some flows form ribbons occupying narrow valley floors; others may gush out in such volume that they bury whole landscapes. A single lava-flow may erupt for days, as is common in Iceland, or weeks, as on Etna, where half the emissions, nevertheless, last less than 25 days. Eruptions lasting for several months are rare. Movement, cooling and solidification occur simultaneously on different parts of a flow, and the rate of discharge and the speed of cooling largely determine its length and surface appearance. Thus, the longest flows are emitted in great volumes at such speed and at such high temperatures that they can travel long distances in a fluid state before cooling and solidification induce deceleration. The shortest flows are extruded slowly, and relatively coolly, and are soon brought to a halt by rapid solidification. Generalizations are hard to make, however, even among similar basalts. Etna, the most diligent purveyor of basaltic flows of any volcano in Europe during historic time, erupts its lavas, on average, at rates of $25\,m^3$ per second, forming flows reaching 7.5km in length and 12m in thickness. However, several Icelandic flows, and many on the Columbia plateaux, are over 100km long, but the longest yet discovered in Auvergne travelled only 50km down the River Alagnon valley. Basaltic flows rarely exceed $1\,km^3$ in volume, but the Roza basalts on the Columbia plateaux cover $40\,000\,km^2$. On the other hand, some lavas are so viscous that they pile up near their vent and cover less than $1\,km^2$. The rhyolites of the Pietre Cotte flow, for example, solidified in a chunky tongue, 1km long, on the 20° slopes of the Fossa cone of Vulcano in 1739.

At first sight, the movement of molten lava seems to be a relatively simple process, but it is governed by many complex factors springing from the properties of the lava itself, the way it flows, and the relief of the landscape invaded. The chemical and physical nature, relative crystallinity, shearing and yield strength, temperature, fluidity or viscosity, and the volatile content all figure amongst the properties of the lava itself. The flow is dominated by discharge rates and volumes emitted, the amount of lava deformation, stirring, and differential cooling, and also by gravity. Valleys then offer the flows their major thoroughfares or restrict them to certain directions: most flows erupted on Piton de la Fournaise in Réunion are confined within the walls of the Enclos Fouqué caldera; and the apparently protecting arm of Monte Somma ensures that most of the flows from Vesuvius are turned southwards to the towns along the Bay of Naples.

Basaltic and silicic lava-flows

The greatest contrasts in behaviour are shown between fluid (mainly basaltic) flows and viscous (mainly silicic) flows. High temperatures, fluidity, high rates of effusion, and copious emissions are often closely linked in the formation of many long basaltic flows. Cool, viscous lavas emitted in small volumes at low rates characterize silicic flows.

Hot, fluid, bright-orange basaltic lavas, surging from the vent at temperatures of about 1200°C, have a striking visual impact. Near the vent, the flow can travel as fast as 20–75 km an hour, but deceleration supervenes when volatile discharges and contact with the cooler atmosphere and the land surface lower its temperature. Cooling also facilitates crystal growth and increases yield strength and viscosity, which slow down the flow still further. The initial orange colour deepens to dark red, first on the edges of the flow and on the crests of the surface waves formed as the lava gushes forth. Larger and larger slabs of cooled, solid, black lava soon appear, floating like rafts on the current. The flow decelerates as a crust of increasing thickness congeals along and builds up its margins. Soon the molten stream is confined to a central axis within its own solidified walls, which *either* takes the form of an open channel like a river, *or* of a lava tube, roofed over and encased completely by consolidated crust. Thus, the main impulse of the current is directed forwards in tongues, rather than sideways, even where the lavas are not restricted to valleys. Only the most fluid lavas, discharged at great rates without forming tubes or axial channels, usually spread laterally to any notable degree. One result of this forward impulse is that basaltic flows in particular quickly approach their maximum lengths and make only slow progress thereafter. It would not be unreasonable, for example, to expect a basaltic flow to reach three-quarters of its eventual length after only a quarter of the emission period had elapsed.

Silicic and some intermediate lavas, on the other hand, behave differently, primarily because they are more viscous than basalts. They are often rhyolitic, dacitic or andesitic in character. These lavas emerge at lower temperatures, broadly around 900°C, which are reflected in their deeper vermilion colour. They are more viscous than basalts because they contain a higher proportion of silica and aluminium dioxide, because they are more polymerized, and because they still initially include much volatile material. Silicic lavas move much more slowly than their basaltic counterparts and rarely reach speeds of 10 km an hour, even just after emission. They thus start off cooler, cool down further, and consolidate before they have time to travel far from their vents. Silicic lavas, therefore, form short, thick, stubby flows, commonly no more than 1 km or 2 km in length. They are also far less numerous and are emitted in much smaller volumes than basaltic flows.

Trachytic, phonolitic and some andesitic lavas often betray less consistent behaviour. They sometimes produce short, viscous flows and even domes, or lava tongues over 5 km long like the phonolites of Teide in Tenerife. A lesser degree of polymeric bonding may play a significant rôle, but it is, perhaps, the different emission temperatures that best account for their variable behaviour.

136

Because some parts of all flows solidify when others are still molten, the motion of the current continually breaks or bends the consolidated crust, thereby creating many distinctive features, while escaping gases form innumerable holes that greatly roughen its upper layers. At length, after further cooling and deceleration, a reddish glow can be detected only at night from a dark, fuming crust that often smells of hot iron. The flow eventually comes to its destination as a grim, cold, brittle, black, grey or reddish-brown tongue, contrasting starkly with the landscape it has invaded.

The behaviour of lava-flows is also influenced by the land, most obviously when valleys confine them to narrow channels, but also when natural obstacles or changes in slope affect their progress. Obstructions often oblige lava that would have lengthened the tongue either to run over an earlier part of the flow and thicken it or to be diverted into branches breaking from its sides. Two recent eruptions of Etna, of similar basaltic composition, illustrate the influence of slope changes. In 1981, a 40-hour eruption with an average discharge of $128m^3$ per second quickly formed a single flow. In 1983, an eruption lasting 131 days emitted five times the volume of similar lava at an average rate of $7–8m^3$ per second and formed a flow with many branches. Yet, both flows reached a length of 7.5km. However, if the 1983 flow had not branched so much, it would have travelled farther than that in 1981. The flow in 1983 branched where the current was blocked, impeded, or changed direction because the slopes became gentler.

Lava columns

When lava cools, it shrinks. As a result, fluid flows often develop distinctive internal patterns, but those in viscous flows are much less regular. The most clearly marked patterns form inside fluid basalts when the current halts before solidification starts (Fig. 7.1). Shrinkage cracks are propagated throughout the lava core, forming polygonal columns, usually with five, six or seven sides, at right angles to the cooling surface (Fig. 7.2). The columns can be over 10m high and 60cm thick, which, when exposed by erosion or quarrying, often show colonnades resembling organ-pipes, such as those at St Flour and Bort-les-Orgues in Auvergne. Practically every valley cut into the Columbia plateaux, notably including the Columbia River gorge itself near the town of Hood River, is bordered by a basaltic colonnade. The columns often stand above a layer of clinkery lava, formed when rubble tumbled from the advancing snout and was buried as the flow progressed. However, the colonnades created when stationary basaltic sheets cooled can extend to the base of the flow, and a thin layer of the underlying rocks is then sometimes baked to a reddish colour by the hot lavas. The colonnade is surmounted by an entablature, often some 10m thick, which is usually made up of thinner prisms arranged in a pattern of fans, stacks or chevrons (Fig. 7.3). They frequently resemble a forest blown down in a gale, like those displayed alongside the Grand Coulee valley in Washington State. Some flows are surfaced by about 3m of rubble, but others are much smoother, and their upper layers are often broken by horizontal joints into stacked grey plates that are traditionally used for roofing in volcanic areas.

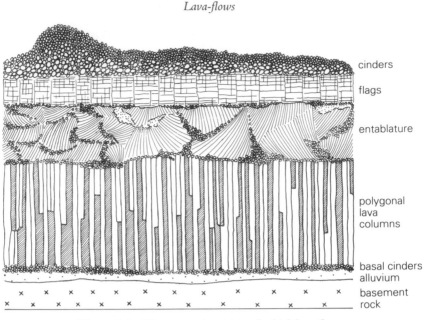

cinders

flags

entablature

polygonal
lava
columns

basal cinders
alluvium

basement
rock

Figure 7.1 Diagrammatic section of a thick lava-flow.

In contrast, the interiors of viscous flows are dominated by shearing and flow structures, caused by slow movements which persist up to solidification, and they generate an irregular maze of cracks, separating small sheets and blocks of different sizes.

Features of lava-flows

The surfaces of lava-flows often reflect their internal movement in relation to their congealing crust. Although there are intermediate types, in broad terms, lava-flow surfaces are either smooth and continuous, or they are notably discontinuous and rugged (Fig. 7.4). The causes of these differences are complex and still incompletely understood. Continuous surfaces form sheet flows and pahoehoe flows. Discontinuous surfaces are subdivided into aa and block flows. The terms "pahoehoe" and "aa" come from Hawaii, and are at least bizarre enough to be unforgettable. Both sheet and pahoehoe surfaces are formed by hot, fluid, less polymerized lavas when the solidifying crust deforms continuously in response to the stresses imposed by the molten current, and develops a continuous, billowing surface. Both aa and block flows are formed by cooler, more viscous, more polymerized lavas, where the stresses generated by the molten current are greater than the solidifying crust can usually withstand, so that it often fails and breaks into innumerable jagged clumps. Sheet and pahoehoe lavas are basaltic. Aa and block lavas may be composed of viscous basalts, but also of andesites, dacites, rhyolites, trachytes and phonolites.

138

Figure 7.2 Lava columns: Picture Gorge, Oregon (overlying John Day Formation ash-flow and older basalts).

Figure 7.3   Splayed lava prisms, Mount Baker, Washington.

Figure 7.4   Contrasting lava-flow surfaces, Caldera de las Cañadas, Tenerife (pahoehoe surface on the left; aa surface on the right).

Sheet lavas

Sheet lavas flood out so fast that vast volumes of fluid basalt are discharged over an enormous area before cooling can begin to halt their progress. Such sheets are the main constituents of the great basaltic plateaux (Fig. 6.5). They emerge from fissure systems sometimes reaching over 50km in length, fed by dykes averaging 8m in width, and form flows commonly ranging between 10m and 30m in thickness. No remotely comparable eruptions have been registered in historic times. Individual sheets have been identified on the Columbia plateau, for instance, which are more than a hundred times more voluminous than any erupted in recorded history. Some 14 million years ago, the Roza eruption in Oregon emitted 1500km³ of basaltic sheets in about a week. In comparison, the largest historical basaltic eruption, from Lakagígar in Iceland in 1783, produced a puny 12.3km³ of lava in eight months. The successive sheets erupt in gushing floods and stagnate before much crystallization and solidification takes place. They form almost featureless plateaux, like those stretching around Pendleton in Oregon, where they overwhelmed the previous landscape so effectively that the Blue Mountain Ridge was the only major upland within the area to escape being swamped.

Pahoehoe lavas

Pahoehoe lava-flows are characteristically thin and smooth, and their relative relief rarely even approaches 10m and is commonly less than 1m. Pahoehoe flows are often formed by small volumes of hot, fluid basalt. Generally, the higher the volume emitted, the faster the current, but except within, say, 2km of the vent, a reasonably agile human can outstrip most advancing pahoehoe lavas. Gas escapes from the mobile surface, and also forms many bubbles, making round holes in the solidified lava. Pahoehoe flows move forwards in tongues or lobes, commonly less than 1m in thickness, which often overlap as they roll onwards. They are coated with a thin, glassy, but still plastic skin which stretches and bulges like an inflating balloon. Molten lava often breaks through the distended skins, spurting out toes, perhaps 50cm or 1m across, which are broken in turn to create yet more toes which solidify to form features the size and shape of an elephant's legs. Thus, the snout of a pahoehoe flow advances intermittently and embraces obstacles in little surges, perhaps up to 5m wide, along a front some 100m broad, at a rate of about 50m an hour.

Lava tubes

Meanwhile, molten lava reaches the advancing snout in a tube (Fig. 7.5). Because solidified basalt is a good insulator, the hot lava in the tube can remain mobile and travel far without suffering a great loss of heat. Hence pahoehoe lava-flows are often long and thin. The tubes join into major, often interlaced trunks, and break up again down stream into distributaries feeding the advancing snout. At the height of the discharge, the tubes are full of mobile lava, but when the discharge wanes, the lava

Figure 7.5 Lava tube on Etna.

often drains away and leaves behind a hollow tube (Fig. 7.6). Lava stalactites, 50 cm
or more long, may form as molten lava drips from the roof, and vertically ribbed
pilasters may coat the tube walls. The current may mark its various levels by forming
solidified benches alongside the tube, like a tide mark on a harbour wall (Fig. 7.7).
Small lava driblets, 30 cm long, then drip from, and solidify below, the bench, like
icicles from a window-sill. The last current may floor the tube with slabs as it congeals, but its regularity can be broken by lava stalagmites, about 50 cm high, if molten lava continues to drip from the roof. Some flows develop an intricate maze of
tubes ranging up to 10 km in length and 30 m in width, although many are much
smaller. Many tubes have been revealed when their roofs have collapsed. In Lanzarote, they are known as "Jameos", and are commonly 5 km in length, and one from
Montaña Corona is wide enough to have been converted into a night club. The
tubes in the Lava Beds National Monument in California have suffered a less exotic
fate. The interconnected systems, generated 11 000 years ago by an eruption from
Mammoth Crater, are sometimes over 16 km long, 30 m wide and 5 m high. As their
names often reveal, a fine array of smaller features decorates many tubes: "Laby-
rinth", a series of interlocking tubes 3 km long; "Balcony", with distinctive side-
benches; "Sleeping Beauty", with rafted blocks later coated by once molten lava;
"Natural Bridge", formed by two superimposed channels; "Mushpot", where new
molten lava broke through the floor of an old tube; "Catacombs", roofed with driblets and walled with ribs that dripped down its sides and now resemble shrouds;
"Elephant Hide Room" and "Flagstone Passage" are self-explanatory; and "Fat Per-
son's Misery" – in this context – represents a constriction narrowing to 15 cm.

142

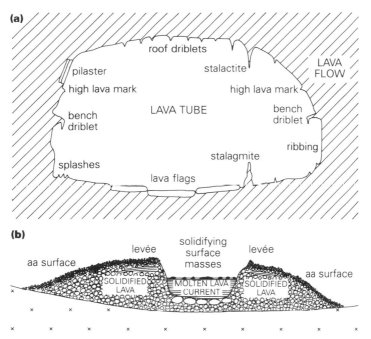

Figure 7.6 (a) Principal features that may develop in a lava tube. (b) Section of an aa lava-flow.

Figure 7.7 Lava tube (not quite roofed over) and driblets, Lanzarote.

Surface features of pahoehoe flows

If they advance unimpeded, pahoehoe surfaces are often pushed into sleek, broad, billowing folds which commonly develop a characteristic bluish sheen when the glassy surface crystals reflect the sky. Smooth surfaces are often covered with filaments resembling an artist's brushwork, some fine, like a Gainsborough painting, others with the rougher texture of a late Rembrandt. If the flow is impeded, pahoehoe lavas develop festoons like coiled ropes, 1–5 cm in height and 2 m in length, which bulge down slope like ridges on hot moving tar (Fig. 7.8). The Blue Dragon Flow in the Craters of the Moon displays these features well. Occasionally, lava is squirted out in a tangled mass resembling entrails, which are well developed, for example, on the northeastern side of the Craters of the Moon. When the molten lava exerts localized pressures on its crust, the surface is raised into "tumuli", humps 2 m or 3 m high and 20 m broad, which sometimes crack open at their crests. Cracked tumuli compose "Captain Jack's Stronghold" at the snout of the Mammoth Crater flows in the Lava Beds National Monument, named after the Modoc chief who held off US cavalrymen for five months in 1873 by making full use of the labyrinthine natural fortress.

Upward surges as the current progresses may also generate pressure ridges which commonly rise about 10 m above the general level and may be up to 1 km in length. They usually occur in two or three festoons trending parallel to the margins of the flow. The zones around tumuli and pressure ridges are often cracked into slabby

Figure 7.8 Festoons of coiled, corded lava on a pahoehoe surface, Pico Partido, Lanzarote (courtesy Juan Carlos Carracedo).

144

pahoehoe surfaces, which are well developed at Yaiza in the lavas erupted in 1730 in Lanzarote. Small emissions of molten lava may escape from the cracks onto the surface and form toes rarely exceeding 1 m in length, which can be notably grooved when they are extruded like toothpaste from a jagged orifice (Fig. 7.9).

Figure 7.9   Lava toes and small cracked tumuli of slabby pahoehoe, Craters of the Moon, Idaho.

Lava lakes

Pahoehoe surfaces also feature on lava lakes (Fig. 6.34). *Primary* lakes form when basaltic magma wells up in a pit or crater, such as those of Kilauea Iki in Hawaii and the Bory and Dolomieu pits on Piton de la Fournaise. An old lava lake also features on the slopes of Pico Partido in Lanzarote, its surface congealed like toffee in a pan, forming swathes of slabby pahoehoe lava. The most notable lava lake in recent times formed in the crater of Nyiragongo in Zaïre from 1927 to 1977, and returned again in 1982. *Secondary* lava lakes form pahoehoe surfaces when flows are ponded back by an obstacle, but little surface relief develops when they solidify. Such a surface was formed when lavas erupted from Pico Alto flooded the floor of the Guilherme Moniz caldera in Terceira about 2100 years ago.

Lava casts

Pahoehoe lava is so plastic that it can embrace trees and mould their trunks. The trunk is often carbonized, leaving behind a lava-moulded cast in the form of a circular pit. Occasionally, lavas mould trees and then partly drain away, leaving a lava-encased trunk standing out like a chimney perhaps 2m high. Lava casts are so varied and so numerous on the northern flanks of Newberry volcano in Oregon that a state park has been set up to display and explain them.

Volatile emissions from lava-flows

Distinctive features develop on flows when volatiles escape from molten lava. Gas bubbles commonly make the surface layers of many lava-flows highly vesicular, but the best-known landforms associated with such emissions are "hornitos", so called because they often resemble the outdoor ovens common throughout the Hispanic world. Mildly exploding gas shatters fluid lava into spatter, which piles up around the explosion pit in steep-sided humps that may reach 10m in height. They closely resemble spatter cones, except that their vents are limited to the depth of the flow. They are common on basaltic flows but are more clearly visible on the smoother surfaces of pahoehoe lavas. Spatter often piles up asymmetrically about a vent and builds up distinctive, cowl-like accumulations. One of the many hornitos in Lanzarote has been christened El Manto de la Virgen because it resembles statues of the Virgin Mary (Fig. 6.32). A famous pair of hornitos, named less imaginatively the Due Pizzi ("two peaks"), were formed on Etna between 1614 and 1624.

Changes from pahoehoe to aa surfaces

Pahoehoe surfaces can change to aa surfaces during a single emission, perhaps mainly because the lavas increase in viscosity as they cool, lose much of their gas, and develop more crystals. Aa lavas never change to pahoehoe lavas, if only because they normally neither heat up nor become more gaseous as they flow. Etna reveals many such transformations, which commonly occur within 1km of the vent, like that on the flow erupted in 1983. At first, thin, fine-textured pahoehoe surfaces solidify, but, as cooling and stirring occur unevenly in the current, rough, knobbly aa surfaces develop down stream that are called "cauliflower aa" because they resemble fields of charred cauliflowers. It is then gradually altered in turn into a characteristic aa surface as the crust thickens down stream.

Aa lavas

Great roughness at all visible scales is the hallmark of solidified aa flows (Fig. 7.10). They are a masochist's paradise. Probably no rock surface on Earth is so rough to touch, so difficult to walk across, or so painful to fall upon. They make a brutal, intimidating landscape. Aa flows have steep margins and rugged surfaces, riddled with holes and deep ragged pits, 3m deep and wide, clogged with cinders, separated by rough clumps, with sharp needles upon spines upon jagged spikes upon craggy pinnacles up to 10m high. It is no surprise that the pioneers in the American West gave them such names as "Devil's Homestead" and "Devil's Garden" and that the

Figure 7.10 Aa lavas erupted between 1730 and 1736 in Lanzarote, with parasol ribbing on the older cinder cone in the background (courtesy Juan Carlos Carracedo).

early settlers in the Azores called them "misterios", or "awesome mysteries".

Aa flows emerge at temperatures most often between 900°C and 1050°C. They are emitted at high rates, often with much lava-fountaining, and are characteristic of viscous magmas such as silicic basalts, andesites, and some phonolites and trachytes. Their remarkably broken surfaces are formed chiefly when their quickly cooled crust is dislocated by the motion of the molten current. Further jaggedness is provoked by exsolved volatiles, which form large almond-shaped holes riddling the upper parts of congealed flows (Fig. 7.11). Although the current may leave the vent at speeds of about 50 km an hour, its margins soon begin to solidify and build up a pair of walls or lava-levées that confine the current to a central axial channel (Fig. 7.6). Extra gushes of lava sometimes spill over the levées and build them still higher. The molten flow becomes concentrated in an open trunk channel, often 10 m wide and deep. Exposure to the open air, however, means that aa lavas cool and slow down more quickly than pahoehoe lavas running in better-insulated tubes. Thus, aa flows usually travel less far and are thicker than pahoehoe flows of equal volumes.

If an advancing pahoehoe flow has a sleek, somewhat reptilian appearance, overwhelming objects like a boa constrictor, an aa flow lurches forwards like a bulldozer. The snout keeps a steep, high wall of black and incandescent, fuming rubble, resembling a huge coal fire needing a stir from the poker. At times, pressure from the advancing molten lava pushes black blocks from the snout, which crash to its base

147

Figure 7.11   Vesicular lava.

and are run over whenever the flow staggers forwards, thus forming the clinkery carpet over which an aa flow progresses like a tank on its caterpillar treads.

Aa flows advance in fits and starts like shuffling, grinding, heaps of rubble, animated with sporadic bursts of energy. They can surge forwards 100m in 10 minutes on a front 100m long, and then halt, fuming and glowing, for hours. But the surges can push down houses, walls and forests. When aa flows finally halt, they may average anything from 3km to 30km in length and from 10m to 30m in thickness, with steep rubbly margins of 35°.

Surface features of aa flows

A certain order may be detected in the apparent chaos of an aa surface. The solidified axial channel is commonly bordered by steep levées forming ridges 5m high, which are composed of thick, vesicular crust topped by rough boulders up to 1m across (Fig. 7.6). Superb levées, for instance, decorate the latest phonolitic flows erupted from Teide in Tenerife (Fig. 7.12). Such levées, or "ribbes", stand 300m apart and rise between 5m and 15m above the axial channel of the Cheire d'Aydat, erupted from the Puys de la Vache and Lassolas.

Occasionally, large masses of solidified crust may be dislodged and rafted along on the current. These "rafted monoliths" can be the size of houses, and they form notable features in the "Devil's Orchard" in the Craters of the Moon, and on the flows erupted from Timanfaya in Lanzarote. Other detached solidified masses may

Figure 7.12 Levées on a phonolitic lava-flow at the foot of Teide, Tenerife.

Figure 7.13 Accretionary lava balls from a phonolitic flow lying on the pumice cone of Montaña Blanca, Caldera de las Cañadas, Tenerife.

be rolled over and over on the current, so that molten lava accumulates around them like snow sticks to a snowball rolled down a hillside. They form "accretionary lava balls", which can exceed 2m in diameter, on the surface of the flow. On steep slopes, however, the balls can gain such momentum that they roll beyond the snout. In a striking example in Tenerife, a steep phonolitic flow on the eastern flanks of Teide has despatched brown accretionary balls across the yellow pumice cone of Montaña Blanca below (Fig. 7.13).

When the internal movements of molten lava continue after the snout has become immobile, loose rubble is often pushed upwards and outwards into curved ridges called lava moraines. The margins of a basaltic flow erupted from Pico Partido in Lanzarote are paralleled by such lava moraines about 5m high.

Block lavas

Block lavas are characterized by individual, often cubic, masses, with relatively smooth faces, and are commonly between 25cm and 1m in size (Fig. 7.14). Block lavas are usually more glassy than aa lavas, and, because they contain relatively few volatiles, their surfaces are much less rough and pitted than aa lavas.

Block surfaces are formed by viscous lavas: many are composed of andesites, dacites and rhyolites, as well as some trachytes, phonolites and silicic basalts (Fig. 7.15). They are most frequently located on strato-volcanoes, where they are commonly emitted after major explosive eruptions, such as that, for example, at Colima in 1975.

Figure 7.14   Block lavas, El Portillo, Caldera de las Cañadas, Tenerife.

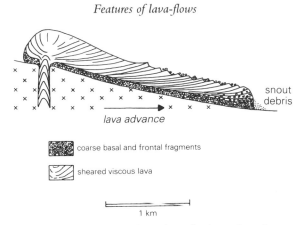

Figure 7.15 Diagrammatic section of a viscous lava-flow.

Because block lavas often emerge at temperatures of about 900°C, and in a fairly viscous state, they tend to be both shorter and thicker than aa flows. They are commonly over 50m thick, 100–500m wide, and can eventually slowly extend for more than 10km in length – although many are very much shorter. Generally speaking, the more silicic the magma, the shorter and stubbier is the flow. Even just after emission, block lavas move only slowly, averaging a mere 1–5m a day in the more silicic varieties. They advance along steep, often lobate, fronts, causing rockfalls and clouds of dust that hide the incandescent mass slowly pushing the snouts forward. The snouts themselves often solidify with slopes of 35°–45° (Fig. 7.16).

The most viscous lavas, notably dacites and rhyolites, well up in bulbous masses that ooze down slope in broad lobes, and it is sometimes hard to distinguish between elongated domes and short block lava-flows – as Pico dos Fragosos in Graciosa amply demonstrates. Block flows of rhyolitic obsidian are notably viscous, especially

Figure 7.16 The snout of Big Obsidian Flow, Newberry caldera, Oregon.

151

when they develop a glassy crust, as is shown by the Rocche Rosse flow on Lipari.

The viscosity of silicic lavas encourages shearing in the plastic interior of a flow, which separates into thin sheets, parallel to the underlying land surface, which slide over each other like slippery new playing cards (Fig. 7.15). The sheets bend and shear upwards near the snout and help extrude blocks that crash down the lava front, or are pushed up from the surface like toothpaste expelled from a tube.

More regular extrusions form curved pressure ridges, or ogives, which often occur in successive waves, over 10 m high and spaced 20–50 m apart, bulging out in the direction of the flow. They are probably caused when spurts of lava increase the internal shearing of the plastic core layers. The ridges show that the whole flow moves as a unit if the lava is viscous enough, and that the differences between crust and plastic core in rhyolitic flows are much smaller than in congealing pahoehoe lavas. Such ridges are very clear on Big Obsidian Flow, which erupted about 1300 years ago in Newberry caldera in Oregon.

Land slope and lava-flows

The slopes of the land often influence the shapes of flows, especially if only modest volumes of lava are emitted. Flows enter valleys, mould their floors, and often cascade over cliffs like black waterfalls, as occurs at Santa Cruz in southern Pico. Perhaps the most famous of all cases where lava travels down steep slopes occurs on the Sciara del Fuoco on Stromboli. Basalts erupt from the main craters of Stromboli at its summit, or occasionally break out from the top of the Sciara itself. Because the slope is covered with unstable screes at an angle of about 35°, the molten lavas quickly slide down hill and often disintegrate before they reach the sea some 700 m below the crest of the Sciara. In contrast, flows tend to branch into distributary tongues when they cross gentle slopes, such as coastal plains or caldera floors. Multiple eruptions develop many coalescent, or even overlapping, tongues, which combine together to create extensive lobes, such as the El Tabonal Negro lava plateau in the Caldera de las Cañadas in Tenerife.

Coastal lava-flows

Varied forms are generated when lavas enter the sea, chiefly depending on the nature of the flow and the relief. On volcanic islands, such as Hawaii and Réunion, the submarine contours reveal distinct bulges, almost platforms, commonly between 50 m and 250 m below sea level, where the majority of flow fronts have chilled and accumulated. These pahoehoe lavas often slink into the sea with scarcely a stir, probably because their hot skin is insulated by a surface layer of steam. Aa and block flows make much more ado, especially when their snouts consolidate into a broad ramp along the water's edge. Further lavas can then pile up behind the ramp, turn sideways in tongues along the shore, and form a broad delta-like accumulation. In Terceira, the flow at Biscoitos, whose name means "biscuit" and "clinker", provides a graphic example of extraordinarily intricate ruggedness.

152

Malpaís

Many separate eruptions, particularly of voluminous aa flows if repeated at brief intervals, form innumerable individual currents, which, in areas of little relative relief, combine together and override each other in a very complicated pattern. The flows build up vast areas of rugged terrain, "like a black and stormy sea of viscid matter suddenly congealed at the moment of its greatest agitation". They are known, for example, as "malpaís", the "bad land", in the Spanish-speaking world (Fig. 7.17) and as "cheires", or "stony ground", in Auvergne. The eruption of Lanzarote between 1730 and 1736 covered a quarter of the island with a rough malpaís. But such eruptions in historic times pale into insignificance in comparison with the flood basalts expelled during brief episodes of the geological past that constitute the most voluminous of all volcanic emissions, apart from those along the mid-ocean ridges.

Figure 7.17   Malpaís in Lanzarote, with Corazoncillo cone in the background.

153

8

Domes

Domes are formed when lava is too viscous to flow as it emerges from the vent, but accumulates and solidifies there in a bulbous mass. They are amongst the smallest, rarest and yet most striking and easily recognizable of the major landforms produced by lava emissions. Although they are much less common than cinder cones, it has been calculated with admirable precision that 217 domes have been formed in the past 10000 years. They are rarely more than 250 m in height or 1 km broad and have steep convex sides and no true craters. Domes are commonly built by rhyolitic, dacitic or trachytic lavas, extruded at relatively low temperatures of about 900°C or 750°C.

Domes occur in environments ranging from summits such as Lassen Peak, to the craters of such strato-volcanoes as Galunggung, and from calderas such as that of Santorini, to continental plateaux such as those in the Chain of Puys.

They have been associated with Pelean nuées ardentes ever since they featured on Montagne Pelée in 1902. Domes usually grow quickly and can be destroyed even faster, especially when they act as corks sealing the vents of strato-volcanoes. In this vulnerable position, the first act of a new eruption shatters them asunder. The dome formed, for instance, in the crater of Galunggung in Java after the eruption of 1822 was completely destroyed by the explosions of 1894; and a further dome, 500 m wide and 85 m high, extruded after the eruption in 1918, was then badly damaged by the eruption in 1982.

In their finest form, they look like the dome of St Peter's in Rome, but they often have a more prosaic resemblance to upturned cauldrons, or, on occasions, to narrow fingers or formless humps (Fig. 8.1). In extreme cases, for instance, the molten lava is so rigid that it is thrust into the air like a piston no wider than the vent, exuding pitons such as Puy Chopine in the Chain of Puys and the sucs like the Gerbier de Jonc in Velay in France. At the opposite extreme, Glass Buttes, just west of Burns in Oregon, is an amorphous spreading mass of beautiful red-streaked black obsidian that is a collector's item. All domes are formed in the open air and should not be confused with much older lava-plugged vents revealed when a volcano is eroded.

The shape of most domes is determined by the way they grow and solidify. They grow from below when viscous lava wells up the vent and they solidify from their outer layers inwards. As the dome forms, then, there is a conflict between the upward-surging, plastic mass and the solid, outer shell of brittle rock preventing its

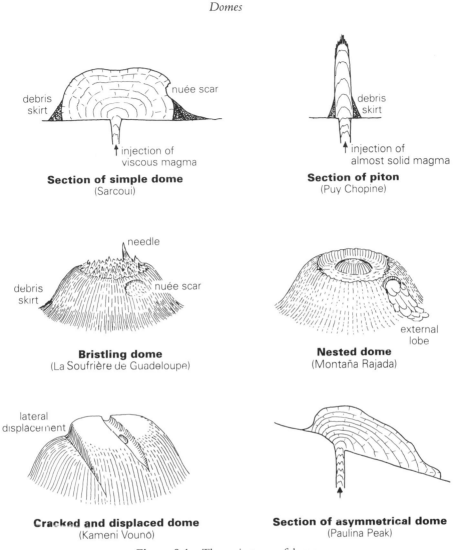

Figure 8.1 The main types of dome.

expansion. Great upsurges can overcome the tensile strength of the outer shell and cause its crest or flanks to burst open, and exude molten material, often along radial cracks. At the same time, gas is often released, occasionally with small nuées ardentes. Ultimately then, the appearance of a dome depends on the interplay of upwelling, explosion and solidification.

Pitons are the simplest of extruded lava landforms. Puy Chopine in the Chain of Puys illustrates their formation. It rises 160m above its base and was formed about 8500 years ago when an almost solid piston of trachyte rose from the vent and was too viscous to spread sideways. The lavas carried upwards blocks of the adjacent country rocks that now cap the piton. Rather more complex clustered pitons develop from closely spaced vents. Black Butte is a sheaf of five dacitic pitons some 9500 years old near Shasta in California, and Roche Sanadoire is composed of five adjoining pitons of phonolite extruded about 2.1 million years ago on the northern flanks of Mont-Dore (Fig. 8.2). The Pitons du Carbet in Martinique represent a looser grouping of six large and four smaller pitons, all clearly separated from each other, that erupted at intervals about 800000 years ago.

Figure 8.2 Roche Tuilière, single piton (left), and Roche Sanadoire, five clustered pitons (right), Mont-Dore, Auvergne.

Sarcoui in the Chain of Puys provides a good example of a simple bald dome with smooth outlines formed by one continuous emission of viscous trachyte lavas about 10000 years ago (Fig. 8.1). It is just like an upturned cauldron, from which its name is derived in the local patois. It has a shallow summit hollow which was probably caused by radial spreading as the lavas settled like jelly tipped from a mould. Its steep sides are surrounded by a skirt of fragments shaken in a semi-solid state from the outer slopes of the dome that helped to support the cooling mass as it swelled up. Quarries also reveal the internal concentric layers of lava, arranged like the inside of an onion. Small nuées ardentes accompanying the growth of Sarcoui form scars on its flanks and an undulating blanket of ash at its base.

More varied relief is created when lava or gas continues to rise up the vent after the shell has solidified. The lava pushes up, stretches and eventually breaks the shell so that molten blocks and pinnacles are thrust up from the main mass to form a rugged, bristling surface that is, in fact, much more common than the smooth outlines of Sarcoui. These protrusions often have a roughly concentric pattern which shows up well, for instance, on the Novarupta dome, formed in 1912 at the foot of Katmaï. In 1902 and 1903, Montagne Pelée developed many needles of incandescent lava rising up to 40m above the dome. The largest, the Needle of Pelée, pointed a sinister finger above the ruins of St Pierre for many months. These needles were pushed up like semi-molten pistons, but they quickly crumbled on exposure to the atmosphere and, in the end, added little to the ruggedness of the summit, which has relief features of little more than 2m on its surface.

The West Indian rival of Montagne Pelée, the andesite dome of the Soufrière in Guadeloupe, has more varied relief (Fig. 8.3). It was probably formed about the time that Columbus reached the island in 1493, and now rises 400m from its base, with a diameter of 900m and a skirt of volcanic fragments. Its surface has been modified by phreatic explosions, which have given the Soufrière some of the most interesting landforms of its type in the world. Cupolas and rough pits abound; rugged pinnacles rise as much as 10m above its surface, and water from 8m of annual rainfall has collected in a small lake. The summit is made all the more attractive by the brightly coloured vegetation that clings to the rocks. The Soufrière is cracked by small faults trending north-south, which have caused its major features. The Fente du Nord and the Eboulement Faujas form two major clefts, more than 10m deep, that gash the outer shell of its northern flanks from top to bottom. They are prolonged across the summit by a line of deep, steep-sided pits blown out by phreatic explosions. One of these, Tarissan, contains a muddy, sulphurous pond; another, wrongly called South Crater, is a dark echoing pit, perhaps 400m deep, still belching clouds of steam every few minutes. These were the main sites of the last phreatic eruption in 1976. One of the smaller clefts on the south flanks emitted a small hot lahar at the start of the eruption on 8 July 1976.

One of the largest domes in the world, 1km across and 500m high, forms the summit of Lassen Peak (Fig. 8.4). About 11 000 years ago, pinkish-grey dacites were extruded and they continually shattered the congealing carapace so that mobile dacite squeezed through the cracks and formed distinct spines and pitons that now make stocky buttresses on the flanks of the dome. The summit has four pits, the largest of which was 300m across and 100m deep when Lassen began its only historically recorded activity on 29 May 1914. This eruption was too weak, however, to destroy the dome, although a year of recurrent phreatic explosions eventually enabled dacites to fill the summit pit and spill over its rim in two tongues on 19 May

Figure 8.3   The summit of the Soufrière dome, Guadeloupe.

1915. The western tongue still forms a black mass of glassy dacite 300m long that solidified on slopes of 20°. The flow that spilled down the steep northeastern flanks was unable to retain its cohesion, and mixed with snow and surface debris to form a lahar that transported 15 tonne blocks more than 30km. A larger eruption followed on 22 May and unleashed a Pelean blast or debris avalanche that devastated a wide area to the northeast of the dome and ripped out the shallow alcove that is still visible on its flanks. About 1650, similar events had also affected Chaos Crags, the dacitic domes some 500m high, which were extruded about 1000 or 1200 years ago at the foot of Lassen Peak.

When slightly less viscous lava is erupted, or when the vent is situated on a steep slope, then the shape of the dome may be flattened or elongated down slope under the influence of gravity (Fig. 8.1). Paulina Peak dome, the summit of Newberry volcano in Oregon, was extruded about 240000 years ago, and now forms a mass of rhyolitic obsidian, 1500m wide, which oozed 4.5km down the gentle outer slopes of the volcano. The Roques Blancos dome on the northwestern slopes of Teide in Tenerife stretches for more than 1km, and on the Greek island of Nisyros the unnamed southernmost dome sends out a bristling arm of lava with nearly vertical sides to form a large promontory jutting out to sea. Santiaguito, the dome that has been growing on the flanks of Santa Maria in Guatemala since 1922, has stretched so far down slope that one side resembles a tongue like a lava-flow, which emits small nuées ardentes. Regional slopes are rarely so similar that viscous lavas can spread equally in all directions. This occurred, however, about 780000 years ago on the plains of Oregon to produce the splayed, symmetrical shape of China Hat.

Figure 8.4 Lassen Dome, California, with protrusions and tongues; the area devastated in May 1915 is in the foreground.

The land surface nowhere has a more spectacular influence on dome development and modification than on the summit of Merapi in Indonesia. The dome occupies an unstable, unsupported position at the western edge of the crest of the volcano and thus cannot maintain its western sector, which is periodically sliced off and cascades down slope, accompanied by nuées ardentes (Fig. 5.11). The Methana peninsula in Greece is composed of many coalescent dacitic domes, some of which have also extended tongues down slope under the influence of gravity (Fig. 8.1). The most recently formed dome of Kameni Vouno, "Burnt Hill", sagged down to the west to make two distinct parts. New lava then welled up from the fracture and formed a rugged lobe at Kameni Xora, which looks so fresh that it was probably extruded during the eruption of about 250 BC, noted by the Greek geographer Strabo. "In Methana", he wrote, "a mountain 7 stadia high was raised up by a fiery eruption. It was unapproachable in daytime because of the heat and smell of sulphur, and at night it shone for a great distance. It was so hot that the sea boiled for 5 stadia around and was turbid for as much as 20 stadia."

Occasionally quite violent explosions affect a dome after the outer shell has solidified. Montaña Rajada ("split") at the foot of Teide had two such explosions (Fig. 8.1). Its summit is marked by a large pumice-filled hollow with jagged sides, caused by an explosion that distributed pumice and shattered domal fragments beyond Montaña Rajada. A smaller dome of similar bristling reddish phonolite was then intruded into its floor, and now looks like a cake set in a jagged bowl. The western flank of Montaña Rajada also burst open and a small, viscous lava-flow escaped to form the latest addition to the dome, and solidified before it reached the foot of its 20° slopes. This lobe shows that domes can grow by external accumulation as well as by internal insertion of lava.

A remarkable example of a two-stage dome development has only recently been recognized in Puy de Dôme in the Chain of Puys (Fig. 8.5). It is the symbol of Auvergne, rising over 200m higher than its neighbours, and a landmark for 100km around, and it has achieved the supreme accolade of often being selected as a finishing point of a stage of the Tour de France. This majestic mountain is larger than most domes, rising 500m above a base that is more than 1000m broad. Puy de Dôme is, in fact, two trachyte domes lying side-by-side with the younger eastern dome partly enclosed by the western. The original eruption created a large bristling dome, like the Soufrière in Guadeloupe, cracked by radiating fissures and pinnacles that still give the rugged outlines to the western flanks of Puy de Dôme. The eastern half was destroyed by a violent blast that formed a deep, east-facing hollow like a Greek theatre or odeon. A second dome, resembling those of Montagne Pelée, then arose within this hollow. Its sides are smooth but its summit is more rugged, possibly where Pelean needles were extruded. All these eruptions probably took place in rapid succession about 10000 years ago.

One of the greatest concentrations of pitons and domes in Europe occurs in eastern Velay in France, where they are known as "sucs" (Fig. 8.6). More than 50 phonolitic extrusions, with some of trachyte, occurred during a brief episode between 7.5 million and 6 million years ago. The sucs dominate the surrounding basaltic

159

Figure 8.5   Puy de Dôme from the southeast. The bulge in the left profile represents the remains of the first, western dome; the summit and right-hand slopes belong to the newer eastern dome.

plateaux. They were extruded relatively calmly, probably without the explosions which are frequent elsewhere. Their steep flanks often display lava columns that commonly break up into plates, known locally as "lauzes", which are widely used in the area for roofing, and which, incidentally, when struck, produce the characteristic ringing sound that gave phonolite its name. The pitons were formed when viscous phonolites were forced out of the vent at temperatures scarcely exceeding 750°C. They probably emerged with the consistency of just-molten glass and often rose to a height of 100 m. The pitons usually have steep sides, often sloping more than 50°, and sharp summits, like their most famous representative, the Gerbier de Jonc, and its essentially similar companions, La Roche Tourte, Le Séponet and Mont Pidgier. True domes were extruded by less viscous phonolites, which perhaps reached temperatures of 850°C on emission. They are well represented east of Le Puy by the Suc de Montfol. La Tortue, which really does resemble a tortoise when seen from Le Puy, was extruded from two separate vents on a small fissure: one producing a low dome forming the shell, and the other a piton representing the head of the reptile. The largest phonolitic accumulations of eastern Velay form the massifs of Meygal and Mézenc, which seem to have been emitted from fissure-aligned vents that gave off more phonolite than other vents in the area. They thus form wide humpy plateaux from which smaller sucs were often extruded by later, more viscous emissions. Mont Mézenc, for instance, is a broad plateau, 6 km long and 5 km wide, crowned by a central dome, which was extruded about 7.4 million years ago. These dome-flows differ but little from the thick silicic lava-flows exemplified by Glass Mountain in northern California. Both belong to a continuum of forms, without clear-cut boundaries, ranging from pitons to silicic flows, whose main characteristics are dominated by the viscosity of their lavas.

160

It almost goes without saying that the most studied domes in the world were formed in the caldera of Mount St Helens. All were composed of dacites extruded from a vent about 25 m wide. Two successive domes grew up as soon as the great eruption began to wane: the first, formed in late June 1980, was destroyed by an explosion in late July; the second was formed in early August 1980 and destroyed soon afterwards. The third and still surviving dome began to grow on 18 October 1980 and apparently ended its present eruptive episode on 22 October 1986. Increases took place in 17 short bursts lasting up to 3 weeks, separated by quiet intervals of several months, although the dome grew continuously from February 1983 to February 1984. The longer the quiet interval, the larger the volume of lava erupted during the next period of growth. Frequently, the invading magma bodily displaced part of the flanks of the dome and cracked its summit. In early September 1984, for example, part of the summit area collapsed by 34 m, but a large lobe of molten lava flowed into the new trough, thus partly compensating for the subsidence. The lobes were often about 20–40 m in thickness and up to about 400 m in length. Then, on 23 May 1985, surveys showed cracks radiating from the centre of the dome and, four days later, invading magma pushed the south flank 100 m outwards and formed a trough crossing the summit of the dome that soon became 500 m long, 250 m wide and 58 m deep. When the main episode of formation drew to a close in 1986, the dome was 267 m high and about 1000 m across, and slightly elongated towards the north following the slope of the caldera floor. Throughout the period of its growth, the height of the dome increased consistently with

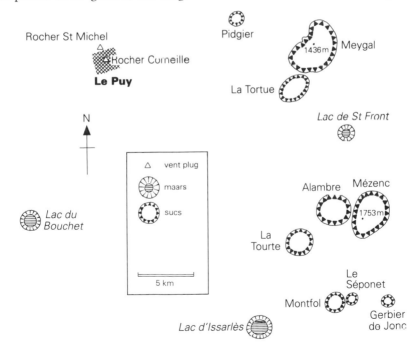

Figure 8.6 Le Puy and selected sucs and maars in Velay, France.

increases in its volume, and the relationship between its height and diameter scarcely altered after June 1981. Before each eruption, more earth-tremors were felt and the dome swelled up slightly; and after each eruption the dome slowly subsided and spread outwards at a rate of between 2mm and 5mm a day as the plastic materials in its core sagged under the influence of gravity. In all, the bulk of the dome had been formed by the insertion of magma from the vent into its core, but about a third of its volume came from lobes expelled onto its surface. Since the dome-forming episode waned in 1986, occasional phreatic explosions have blown pits into its surface when water percolated to the still-hot core below.

These studies at Mount St Helens provide valuable insights into the way domes actually grow. If Mount St Helens is typical, then their development is more complicated than was previously thought. One major problem about studying older domes is that, unlike cinder cones which are often quarried or eroded, sections are mostly unobtainable, and a simple outline can thus mask intricate structures caused by complex growth. It has been acknowledged, for instance, that it would be hard to reconstruct the detailed story of the growth of the Mount St Helens dome from the features now displayed on its surface. Many distinctive surface features of several domes already described also formed at Mount St Helens. For example, the lobes formed on Mount St Helens resemble those at Kameni Vouno and Montaña Rajada; the trough is a larger version of the clefts on the Soufrière dome in Guadeloupe; and the sagging of its flanks resembles that at Kameni Vouno. Consequently, the features apparently distinguishing many domes may be a matter of chance, rather than showing a truly different origin.

9

Cinder cones

Cinder cones are among the most common volcanic landforms found in the world, but very few are famous, many have no name, and, as their eruptions rarely cause any loss of life, they scarcely ever make the headlines. They are also amongst the simplest products of volcanic action and, although many cinder cones are as alike as peas in a pod, they often show up clear differences of both eruptions and landforms which make it easier to understand more complicated volcanic processes (Fig. 9.1).

Cinder cones are chiefly produced by Strombolian eruptions. They commonly grow up in groups and are often found on fissures or in swarms, in both oceanic environments such as Iceland and continental environments such as the Craters of the Moon or the Chain of Puys (Figs 9.2, 9.3). They also often occur on the flanks of strato-volcanoes or shields. For example, the Monti Rossi, erupted in three

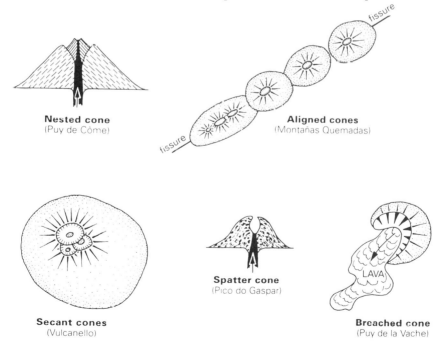

Figure 9.1 The main types of cones.

163

months in 1669, Monte Leone, formed in three days in 1883, and the Monti Silves-tri, accumulated in five months in 1892, are only a selection of the cones which have appeared on the flanks of Etna during the past four centuries (Figs 9.4, 9.5). No less than 400 cones feature on the slopes of Newberry volcano in Oregon. Cinder cones also constitute one of the main ways in which calderas are filled, and, for instance, more than 350 have erupted on the floor of the Enclos Fouqué caldera in Réunion in the past 4745 years. Cinder cones are formed by moderate explosions of basalt or andesite with a moderate gas content. In general, they are composed of coarse, rough fragments, ranging in size from ash to large clinkery cinders, which are

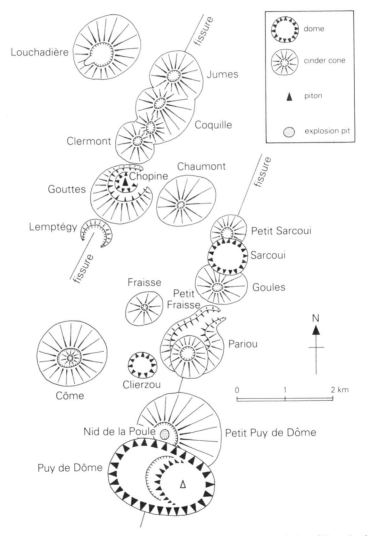

Figure 9.2 The cones and domes in the central part of the Chain of Puys in Auvergne (lava-flows omitted).

Figure 9.3 The Chain of Puys from the southwest.

arranged in beds, usually over 10 cm thick, that are best developed towards the periphery of the cone. A more chaotic mixture falls back and accumulates near the vent and is very well displayed, for example, in the deep craters of Timanfaya and Montaña Rajada, formed during the eruption of Lanzarote between 1730 and 1736 (Fig. 6.31). The bulk of each cone, however, is composed of regular layers of loose fragments of similar size, sloping downwards away from the summit. A layer of ash may be superseded by a layer of cinders, or spatter, or lapilli, followed by another

Figure 9.4 Cinder cones formed between 1730 and 1736 near Islote de Hilario, Timanfaya, Lanzarote (courtesy Juan Carlos Carracedo).

165

Figure 9.5 One of the craters of the Monti Silvestri with another cone of the group, also erupted in 1892 from a small fissure. Montagnola, which erupted on the same fissure in 1763, rises on the horizon.

ash layer and so on. Larger blocks and bombs are sometimes expelled, but rarely in sufficient quantities to form beds of their own (Fig. 9.6). The size of the fragments ejected is determined by the intensity of the explosions which shatter the magma close to the exit of the vent. The fragments cool during their journey through the air, and fall back to the ground in a solid state, accumulating in friable layers lying at their natural angle of rest. Ash lies at about 25°, and cinders at about 32°, so that the smooth slopes of most freshly erupted cinder cones usually average between these angles.

The summit of a cinder cone is marked by a crater, situated directly over the vent, that is commonly 200m across and surrounded by a clear, circular rim. The crater is typically funnel shaped as long as the activity continues, but it becomes bowl shaped, and often between 20m and 50m deep, when lava or fragments plug the vent at the close of the eruption. The crater slopes are determined by the explosions that splay the fragments outwards and upwards like a gigantic pea-shooter. Overhangs do not exist in the crater because of the repetition of blasts during the eruption and the looseness of the fragments forming the crater walls, which are adjusted to their angle of rest. Cinder cones grow rapidly and soon approach their maximum size, however long the eruption continues. They only rarely exceed 250m in height and 500m in diameter. For example, three-quarters of the cones in the Chain of Puys are between 300m and 500m across at their bases and rise to a height of about 100m. Cinder cone eruptions usually end in a year or two; half of those recorded in the past century, for instance, continued for less than a month and 95 per cent were active for less than a year. The only major exceptions appear to be the cinder cones formed on the summits of the more effusive strato-volcanoes. On Etna, for example, the cinder cone of North-East crater was built up over almost 30 years; that of South-East crater has been growing since 1978.

The repetition of eruptions of similar intensity over a short period produces relatively simple forms. The most common result is a simple cinder cone, crowned by a sharply defined crater, with straight slopes of between 25° and 32°, both inside the crater and on the outer slopes of the cone, and one or more lava-flows may issue from its base. Puy des Goules in Auvergne, Lava Butte and Pilot Butte near Bend in Oregon, Bagacina in Terceira, Chinyero in northwestern Tenerife, and Wizard Island in Crater Lake, Oregon, all represent this simple volcanic landform in different environments. Puy des Goules, for instance, is well preserved despite its 31 000 years, for its still ungullied grassy slopes lie at an angle of 25°. It is 175 m high, 900 m across and its bowl-shaped crater is 35 m deep and 200 m wide.

The shapes of cinder cones are often changed by the interplay of small variations during their short lives (Fig. 9.1). They can be modified, for example, if the vent shifts, or if the explosions vary in intensity. When the position of the vent alters (most notably along fissures), aligned, twin or secant cones and craters develop. When the power of the eruptions varies, then nested, buried or breached cones are formed. The sequence of events is therefore often very important in creating the resulting landforms.

Cinder cones erupt on fissures in several ways. Activity may migrate along a

Figure 9.6 Bombs, cinders, ash and pumice; Broken Top, Oregon.

167

fissure during a single eruptive episode, or it may take place throughout its length at the same time, or eruptions may occur on a fissure at different times from separate vents. The 18th-century activity in Lanzarote migrated along a fissure and most of the recent eruptions of Etna have progressed down slope on fissures. The mid-ocean ridge emissions of Iceland, such as that of Lakagígar in 1783, offer the best examples of contemporaneous fissure emissions, where the cones are so closely spaced that their beds are intermingled and their individual outlines indistinct. On the other hand, the fissures in the Chain of Puys operated sporadically (Fig. 9.7). On one fissure, for example, Puy de Lemptégy was formed about 30000 years ago; the twin cones of Puy de Jumes and Puy de la Coquille erupted together, accompanied by the nearby Puy de Clermont, about 10000 years ago; and the complex of Puy Chopine and Puy des Gouttes erupted about 8500 years ago. (The presence of the piton of Puy Chopine in the midst of these cinder cones also demonstrates that a single fissure need not necessarily always erupt in the same style.) Sometimes, too, vents are so closely spaced along one fissure that they form a single cinder cone. Thus, Puy de Barme in the Chain of Puys and Belknap cone at McKenzie Pass in Oregon are both elliptical cones, elongated along their parent fissures and crowned by three aligned summit craters, formed from three closely spaced vents. In an extreme case in Velay, Mont Devès crowns a line of coalescent cones forming a wall 4km long and 150m high that erupted together.

Shifts also occur when vents are clustered together. Sometimes the vents are so

Figure 9.7 Varied eruptions on a fissure in the Chain of Puys. Petit Puy de Dôme cinder cone, with Nid de la Poule phreatic explosion pit on the left foreground; Puy de Pariou cinder cone in the middle distance; Puy des Goules cinder cone just behind, with Sarcoui dome farther behind.

Figure 9.8 Puy de Lassolas (left) and Puy de la Vache (right), Chain of Puys (courtesy Jean–Paul Gadet).

closely spaced that they form a single, bulky cone with intersecting summit craters. This is admirably illustrated by Vulcanello at the northern end of Vulcano, where three craters intersect as a result of shifts of about 100m in the vents used by successive eruptions since 183 BC. The original crater has been damaged the most, whereas that formed by the last eruption in 1550 is naturally the only one to have preserved its steep inner slopes and its perfectly circular rim (Fig. 9.1).

In a single vent, the changing intensity of explosions influences the appearance of all cinder cones. This is where the position of the most violent explosion in the time sequence is important. An unusually large eruption of fragments may completely bury the original cone, which can then be detected only after quarrying or erosion. Puy de Tartaret and Puy de Mercoeur in the Chain of Puys both cover buried cones. In contrast, an explosion of unusual violence can blast out a larger crater which may then be gradually filled in by a new, smaller cone, formed by weaker eruptions. This new cone is cradled or nested within the larger crater (Fig. 9.1). Puy de Côme in the Chain of Puys provides a beautiful example where a more violent explosion occurred when the cone had already reached its full, and greater than usual, height of 350m. A small new cone with its own new crater was then built, which had only partly filled the old widened crater when the eruption stopped. The two crater rims now form concentric ridges at the summit, separated by a circular hollow or atrium, which is a characteristic feature of such nested cones.

Breached cinder cones can originate in several different ways. A breach is a kind of wound that breaks the continuity of the cone and its crater so that their circular outline is incomplete because one sector is missing (Fig. 9.1). Breached cones are thus asymmetrical and are often crescent shaped and very similar in form to a French croissant (a "crescent"; Fig. 9.8). Perhaps the most famous breached cones in Europe are Puy de la Vache and Puy de Lassolas in the Chain of Puys, although Montaña Reventada in Tenerife, whose very name means "breached", deserves to be better known. A cone can be breached when a sector is destroyed by rapid upwelling of lava, or by lateral explosions of unusual violence, or, more rarely, when a large-scale failure such as a landslide removes a sector. Breaching also occurs during an eruption if continued lava emissions carry away any fragments falling upon it and thus stop one sector of the cone from being built. Similarly, a strong prevailing wind may blow most of the fragments to one side of a vent and therefore restrict accumulation on the opposite side.

Rapid upwelling of lava destroyed the western slopes of North-East crater at the summit of Etna early in 1978, when they pushed its flank more than 200m down slope away from the foot of the cone. In the Chain of Puys, basalts surged up within Puy Chalard about 50000 years ago and pushed a sector of the cone some 250m westwards, where it was left stranded in a crescent-shaped mass facing onto the breached crater. At Puy de Louchadière in the Chain of Puys, a less copious upsurge of lava breached the western sector of the crater only, and gave the cone the armchair shape from which its patois name is derived ("Lou Chadière, La Chaise"). Breaches also sometimes occur when lava fountaining along fissures cuts through cinder cones. Pu'u O'o cinder cone was damaged in this fashion as it was being formed on Mauna Loa in Hawaii in 1984. These same fissure eruptions also sliced through already extinct cinder cones and breached them on two opposite sides.

An even more unusual instance of cone-breaching took place at Tinguaton volcano in Lanzarote (Fig. 6.33). It is a low cone, formed during the autumn of 1824. Tinguaton was breached by an eruption of hot water, which eroded a narrow gap about 30m wide in its northeastern side, and deposited a little delta of transported lapilli beyond it. The provenance of this water is unclear, but it could well have been heated and recycled groundwater that emerged in a brief, but vigorous hot spring.

Breaching by unusually violent lateral blasts is relatively rare in cinder cones, because their explosions are normally mild and too weak to destroy a sector. Cones can, however, be damaged by large hydrovolcanic explosions, such as those that formed the Narse d'Espinasse to the detriment of the nearby Puy de l'Enfer in the Chain of Puys.

Many cinder cones in Lanzarote and Fuerteventura in the Canary Islands are breached facing the persistent northeast trade winds. These trade winds may have been strong enough to blow fragments onto the southwestern sides of the cones, while impeding accumulation on their windward, northeastern sides, so that true cones never formed but always remained open to their very bases. When lava-flows were later emitted, they would merely exit by the gap on the northeast. However, in the Chain of Puys, where the prevailing southwesterlies are by no means of trade wind persistence, many cinder cones are slightly higher on their northeastern side, towards which the fragments were blown. Nevertheless, the asymmetry, even on a large cone such as Puy de Côme, is limited to their uppermost 20m or so (Fig. 9.9).

In cases where cinder cones erupt on, or on the brink of, steep slopes, a sector of the cone can collapse down slope under the influence of gravity when lavas surge out. Such breaching is common on strato-volcanoes such as Pico or Etna, for instance, and seems to be the prime cause of the formation of Montaña Reventada in Tenerife. Two particularly clear examples also occur on Timão and Quitadouro on the island of Graciosa, both of which erupted on the brink of steep scarps.

On rare occasions, and under special conditions, cinder cones can be breached by small glaciers. If large glaciers form on a strato-volcano, the unconsolidated fragments of the cones may be quickly removed entirely, but the required delicate balance of glacier and cone size, slope angle and gravity was apparently achieved at Mount Bachelor in Oregon. The northeastern sector of its summit cinder cone has been breached by a little glacier that accumulated in its crater, pushing the fragments down slope *en masse*.

Perhaps the most common explanation for cinder cone breaching is that the explosion of cone-forming fragments takes place at the same time as the emissions of lava. When the lava leaves the vent, any fragments falling upon it are transported away as if they had landed on a conveyor belt. The cone, therefore, never grows above the lavas, but fragments still accumulate on the remaining sectors. Eventually, if the flow continues and the wound never heals, the fragments form a crescent-shaped hill, open on one side to its very base where lava left the vent. Such is the likely origin of Montaña Corona, in northern Lanzarote, whose breached crater is 400m deep; and of the equally deep North Crater in the Craters of the Moon; as well as of those famous breached cones, Puy de la Vache and Puy de Lassolas in the Chain of Puys (Fig. 9.8). They probably never existed as circular cinder cones and should therefore really be described as cinder crescents. This may well be the best explanation, too, for the crescents associated with lava-flows in Lanzarote and Fuerteventura. If upsurging lavas breach a cone after it has been formed, then the displaced mass of fragments is pushed out and piled up alongside the flow. Crescents have no such accumulations, because the fragments falling onto the molten current are simply absorbed into the surface of the flow.

Figure 9.9 Puy de Côme, with small nested cone and asymmetrical profile, Chain of Puys (courtesy Jean–Paul Gadet).

Spatter cones

Sometimes hot erupting magma contains just enough explosive gas to prevent the formation of a lava-flow, but not enough to shatter it into small fragments. In this case, the lava is torn by the expanding gases into hot, fluid clots, or spatter, varying in size from 1 cm to 50 cm across, often thrown into the air in lava-fountains. Spatter ramparts are formed when the lava-fountaining takes place along a fissure, but spatter cones are formed when the eruptions become focused on individual vents. Spatter is distinguished from many other lava fragments because it falls back to Earth in a molten state and is deformed on impact to form flattened masses called cowpats. The clots can weld themselves together as they solidify so that spatter forms steep-sided accumulations, often built up almost vertically like rough walls around the vent (Fig. 9.1). Thus, spatter cones and ramparts are usually much steeper and rougher than smooth-sided cinder cones whose fragments are loose and mobile. The contrast is neatly illustrated in the Lava Beds National Monument in northern California, where the rugged spatter cones of "The Castles" and "Black Crater" rise near the base of the smooth Schonchin cinder cone (Figs 9.10, 9.11).

Spatter cones and ramparts are commonly much smaller than cones of ash or cinders. Spatter ramparts are rarely more than 30 m high because lava-fountaining is a rather ephemeral, though spectacular, aspect of fissure eruptions. It was common, for instance, in almost all the initial outbursts during the eruptions of Krafla in Iceland between 1975 and 1984. The little spatter cones on lava-flows, known as hornitos, also result from short-lived activity, but in spite of their small size they retain

Figure 9.10 Spatter pinnacle, Black Crater, Lava Beds National Monument, California.

172

Figure 9.11 Schonchin cinder cone with smooth slopes, contrasting with the spatter cone at its left base; Lava Beds National Monument.

their typical steep slopes and thus stand out especially clearly, for instance, on the pahoehoe flows at the Craters of the Moon, or in Lanzarote. It is only in the comparably infrequent cases when lava-fountaining is prolonged that spatter can construct masses rivalling cinder cones in size. Such eruptions have given rise to Pico do Gaspar in Terceira, which is more than 100m high (Fig. 9.12), and to the ragged black peak of Pico Partido, formed in 1730, which rises over 200m above the lava-flows in Lanzarote (Fig. 6.30). All spatter cones retain very similar characteristics. Their outer slopes exceed 40°, whereas their inner crater slopes are often vertical or even overhanging, where lava clots thrown from the vent have stuck to the sides of the crater. At times, the clots of spatter drip down and form driblets and, perhaps, lava stalactites, which can join together to form columns or pilasters attached to the crater walls. The crater rim often develops overhangs when the spatter has stuck to the lip, either on its way skywards, or as it falls back to the ground. Once an overhang develops, it becomes more likely to trap further spatter, so that the rim can jut out for as much as 3m over the crater, as occurs on the summit of Pico Partido (Fig. 6.35). As a result of the agglutination of so many clots of spatter, the craters of spatter cones are often shaped like a wineglass, deep, narrow and steep sided, and occasionally with an overhanging rim. Because the spatter is welded, their craters retain their typical steep profiles much longer than those of cinder cones, whose unconsolidated fragments are soon degraded into open, bowl-shaped hollows.

Figure 9.12   Pico do Gaspar spatter cone, Terceira, Azores.

10

Strato-volcanoes

Strato-volcanoes are the stars of the volcanic world. They frequently form imposing cones, often exceeding 2500 m in height, 1000 km^2 in area, and 400 km^3 in volume. Fuji, Orizaba, Mount Rainier, and Kilimanjaro, for instance, range between 500 km^3 and 900 km^3 in volume; Etna, the largest active volcano on land in Europe, covers 1250 km^2, but the now extinct Cantal volcano in Auvergne overshadows it with an area of 2750 km^2. Many strato-volcanoes form impressive, snow-capped peaks. Etna dominates half of Sicily, Popocatépetl stands guard over Mexico City, and Mount Hood signalled the promised land to the 19th-century pioneers well before they reached the end of the Oregon Trail (Fig. 10.1). But other strato-volcanoes are quite small: neither Cosegüina, Montagne Pelée, nor Vesuvius reaches 1500 m. Almost all, however, have a beautiful, often isolated, individuality, reinforced by a threatening presence that is expressed, from time to time, by gran-

Figure 10.1 Mount Hood, Oregon, from the air, with Mount Jefferson in the background on the right.

175

diose – and fortunately rare – eruptions which can wreak some of the most rapid morphological changes on Earth and bring death and destruction in their wake. In between times, strato-volcanoes are often so quiet that they seem extinct – so that to witness the start of a great eruption needs luck or careful surveillance, whereupon even greater care must be taken to stand very well clear.

These spectacular eruptions commonly excite great scientific interest and sometimes add hitherto unknown features to the volcanic repertoire. Pliny the Younger's account of the paroxysm of Vesuvius first described what was to become known as a Plinian eruption, and, appropriately, Vesuvius has been its most diligent European exponent ever since. Montagne Pelée brought nuées ardentes to the attention of the scientific world in 1902, Katmaï revealed rhyolitic floods in 1912, and Mount St Helens highlighted the rôle of blasts and debris avalanches in 1980.

Just as strato-volcanoes stand out in the landscape, so their behaviour dominates the districts around them. Their violence is legendary. But each strato-volcano is a character in its own right, not always behaving with regimental uniformity. Cinder cones, for instance, are monogenetic, almost monomorphic, features, notably limited in their range of eruptive style, length of activity, and landforms. On the other hand, strato-volcanoes are polygenetic, and capable of displaying almost the whole panoply of eruptive styles during "nasty, brutish and short" outbursts, often separated by long interludes of repose, in a total active life lasting many thousands of years. Repeated eruptions from a central focus are the hallmark of strato-volcanoes. Their most frequent and most voluminous eruptions usually take place from a central vent, or cluster of vents, expelling pumice and ash and occasional short, thick, viscous flows. Flank and peripheral eruptions, chiefly of lava-flows, also occur, which may be individually large, but they are less frequent, more scattered and more fluid, and thus thinner than those emanating from the centre.

The result of these spatial variations in activity is the formation of a cone often with a broadly concave profile – whose steep summit slopes wane gradually to a much more gentle periphery. Because the eruptions take place over thousands of years, atmospheric erosion removes materials from its upper areas and deposits them, often in a broad apron, around its flanks – thereby emphasizing its concavity.

These cones add much to the aesthetic quality of a volcanic landscape, but, in fact, their serenely simple outlines often mask a nightmare for geologists trying to disentangle their history. Their complicated biographies are not the least of their fascinations. Strato-volcanoes owe their rather inelegant name to the stratification of many different kinds of erupted fragments and lava-flows. But the succession of layers can be extremely complex. Moreover, especially during periods of repose, erosion excavates deep valleys into the volcano, which then become the almost obligatory paths for many future emissions. The normal geological laws of superposition, whereby newer beds lie above older beds, thus no longer often apply directly because the layers can be almost inextricably mingled. Hence the proper sequence of events is difficult to elucidate, even with the most sophisticated modern techniques. In addition, activity lasts for so long that the most recent eruptions often cover the products of many predecessors. Thus, the history of many strato-volcanoes is often the subject of constant discussion and revision. In fact, older, extinct strato-volcanoes are commonly displayed in more intricate detail – even if they are not always better understood – than their most active companions, because streams have eaten into their ancient cores and exposed at least a selection of the eruptions that gave them birth. The early history of Etna, for example, is incompletely known because its basement is now only partly visible around its southern edges. But a plethora of varied eruptions has been revealed by the deep erosion of the extinct Mont-Dore.

Distribution of strato-volcanoes

Strato-volcanoes are constructed most frequently along subduction zones, where they often form impressive chains on land, and island arcs at sea. There are more than 500 active strato-volcanoes associated with plate edges in and around the Pacific Ocean, and they are also found in, or near, rift zones and sometimes on the flanks of mid-ocean ridges. However, the great oceanic piles – such as Réunion and Hawaii – are not strato-volcanoes but enormous shields formed in the main by prolonged, mild effusions of basaltic lava and, although they often have a simpler structure, they are as much as a hundred times larger than the bulkiest of strato-volcanoes.

Forms of strato-volcanoes

Although strato-volcanoes are frequently large and conical, their detailed forms reflect contrasting histories ranging from constructive emissions and deposition to destructive explosions and erosion. Thus, for example, except for the much-degraded stumps of Mount Washington and Three Fingered Jack (whose photographs were unpatriotically confused in one geological guide to the region), most visitors to the Cascade Range could probably distinguish each volcano after a day or two. If the materials erupted from a strato-volcano can spread unimpeded, they form a circular or elliptical accumulation. A circular plan results from a broadly equal radial distribution from a central focus. An elliptical plan may result from lateral shifts in the positions of the vent, or from an extension along summit fissures. Hekla, for instance, is an unusually elongated strato-volcano because it developed from vents along a summit fissure 5 km long (Figs 10.2, 10.3).

The cross-profiles of strato-volcanoes show greater variation (Fig. 10.2). The smooth, soaring, steeply concave cones represented by Mayon in the Philippines, Agua in Guatemala, and Pavlof in Alaska seem best developed where recent, predominantly explosive, activity has been strongly concentrated on a sharply defined crater above a stable central vent, and where radiating nuées ardentes, lahars, or even snow avalanches help maintain their graceful, concave regularity (Fig. 10.4). Such cones now expel few lava-flows from their summits. In many cases, the conical out-

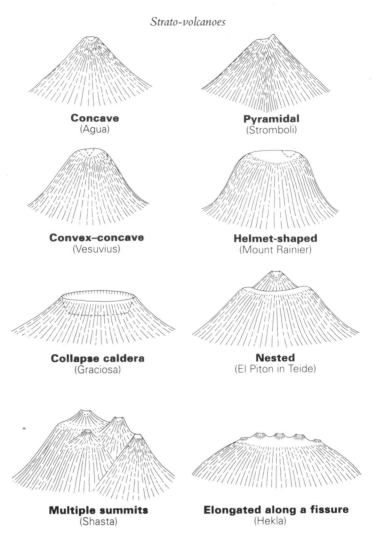

Figure 10.2 The main types of strato-volcano (not to scale).

lines are deformed, most often because the summit vents shift laterally, or because their flanks are faulted. Stromboli, for instance, has a pyramidal shape, first caused when the vent migrated 1 km northwards, and then accentuated when the Sciara del Fuoco sector foundered on its northern side. On other strato-volcanoes, convex upper slopes curve down to lower concave slopes. They seem to have recently emitted more lavas from central vents than their more concave counterparts, often piling up stubby, viscous flows near their summits. Popocatépetl, Pico, Chimborazo and Teide are amongst those with such profiles. Greater summit convexity is displayed by strato-volcanoes with helmet-shaped outlines, exemplified in the Cascade Range by Mount Rainier and by Mount Adams, where several closely spaced summit vents emitted viscous lavas that merged into a common pile. More widely spaced vents

Figure 10.3   Hekla, Iceland (courtesy Margot Watt).

give rise to more distinct summit features. Thus, among the volcanoes of the Cascade Range, Shasta has two clearly separated summits, as well as a pair of older, half-buried predecessors. Other strato-volcanoes are more obviously composite in character, comprising separate accumulations which piled up when activity migrated from one vent focus to another. Mont-Dore volcano is one of these. It grew up from eruptions at four successive centres over a period of some 18 million years. The eruptive centres of Vulcano have migrated northwards (Fig. 2.2). The original strato-volcano emerged in the south; successive calderas collapsed on its northern flanks; Fossa then grew up on its northern periphery; and Vulcanello emerged still

Figure 10.4   San Pedro, Guatemala.

179

farther north. If Vulcano runs true to form, the next eruptions should occur very close to Lipari Island.

Many strato-volcanoes also develop summit calderas which so severely alter their profiles that they merit separate treatment in Chapter 11.

History of strato-volcanoes

Although, in terms of the age of the Earth, all active strato-volcanoes are young, they often figure amongst the oldest of volcanic landscape forms. Many, however, have not yet been accurately dated, so that their true age is uncertain. Nevertheless, many active strato-volcanoes are less than a million, or even 100 000, years old, although, at times, they may rest upon older plinths. Montagne Pelée, for example, is about 35 000 years old, but its volcanic basement is at least ten times older. Shasta is less than 250 000 years old, and most of it formed less than 20 000 years ago, but it lies on the remains of an older sibling, decapitated by a debris avalanche 300 000 years ago. In fact, the conventional image of strato-volcanoes – of notorious cones delivering violent explosions – often only develops late in their lives, but very quickly, as if they had suddenly fallen prey to unbridled passions in their maturity.

Basaltic plinths

Many strato-volcanoes lie on broad plinths constructed by prolonged basaltic emissions. They are sometimes called basal shields, but they differ from shield volcanoes that accumulate from central vents to heights commonly exceeding 3000 m. Although the plinths beneath many strato-volcanoes are widespread, they are often less than 1000 m thick. They are erupted from multitudes of vents, frequently arranged along fissures, that give rise to a rash of cinder cones and abundant basaltic lava-flows (Fig. 10.5). Intermittent eruptions, repeated for a few million years, eventually build up a plinth spreading 40 km in diameter and sloping at an overall angle of less than 10°. The formation of the plinth constitutes the longest eruptive period in the history of the strato-volcano, so that many, in fact, spew out basaltic flows and fragments in great quantity for most of their active lives. Basalt is therefore by far the most common volcanic rock, even in subduction zones. The basaltic plinth of Popocatépetl, for example, developed for a million years, reaching a diameter of 28 km and a volume of 500 km^3. Its Mexican companions, Colima, Orizaba, and Nevado de Toluca, also grew from basaltic foundations of comparable age and size. Etna offers the finest European example of a basaltic plinth, which formed between 700 000 and 100 000 years ago. Its slopes average about 5° and it covers an area of 1250 km^2. In the Cascade Range, where Mount Adams, Shasta, McLoughlin, Thielsen, Jefferson, North Sister and South Sister all rise above basal plinths, basalts are six times as voluminous as the andesites in the cones above them. However, not all strato-volcanoes necessarily grow from such foundations; and, for

180

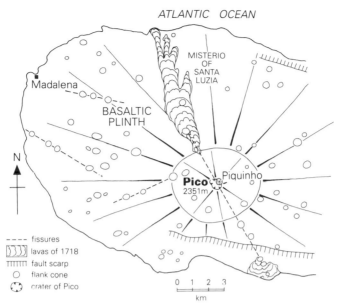

Figure 10.5 Selected features of Pico volcano, Pico Island, Azores.

instance, neither Mount Rainier nor Mount St Helens apparently stand on plinths directly related to their own eruptive centres. Occasionally, the activity building a basal plinth begins to concentrate upon vents arranged in a central focus, without otherwise altering the type of basaltic eruption. Newberry, Medicine Lake volcano, Cantal and Etna all developed hubs with more strongly focused active centres, still emitting basalts, before their eruptive styles changed markedly. The volcanic systems, such as Krafla, in the central rift zone crossing Iceland may also be related to this type of concentration.

Basaltic cones

Basalts sometimes erupt from a concentrated focus for a long period. Short flows are frequently emitted from the central vents, whereas flank eruptions are limited in both volume and incidence. The centre thus grows much faster than the periphery and a basaltic cone results which may eventually achieve great size. Fuji is one of the world's largest volcanoes of this type, and Pico in the Azores, Beerenberg in Jan Mayen, and Stromboli (at least above sea level) provide other examples (Fig. 10.6).

Figure 10.6   Pico volcano, Pico Island, with Piquinho at the summit.

Violent eruptions and periods of repose

Evolution towards violent eruptions seems often to occur when shallow magma reservoirs relay larger, deeper reservoirs in which a range of changes may take place that have already been outlined in the chapter describing the ascent of magma. These transformations have three major interlinked repercussions – changing the nature of the volcanic materials expelled, the character of the eruption, and the appearance of the resulting landforms.

When a strato-volcano adopts a more violent eruptive style, the appearance of its summit area usually alters radically. Most fragments, and any viscous lavas are expelled from a vent, or a central cluster of vents. They build a central cone with slopes commonly ranging from 20° to 35° and rising perhaps between 1000m and 2000m above its base, from which it is usually distinguished by a clear increase in gradient. In some cases, it is so distinct that it is given a separate name. Thus, for example, the upper part of Etna has been called Mongibello.

At the same time, strato-volcanoes then typically expel more evolved magmas. In subduction zones, these commonly include andesites (as on Mount Rainier), dacites (as on Mount St Helens) or rhyolites (as on Katmaï); those on the flanks of mid-ocean ridges and in continental areas commonly erupt trachytes, like Sete Cidades, or even rhyolites, like Mont-Dore. Concomitantly, eruptions become more notably violent and generally brief. However, although Plinian, Pelean or Vulcanian styles are most often represented, almost the whole eruptive range can be displayed during a major outburst lasting only a few days. They are manifested in blankets of ash and pumice that not only form large cones, but may be also discernible in thin layers over 200km from the vent. Nuées ardentes, lahars, blasts, or debris avalanches as well as the collapse of the summit into a caldera may signal the apogee of the eruption. Only the waning phases of the outburst – if at all – usually witness emissions of lava-flows or domes. These climactic phases are episodic and short-lived: the climaxes of Tambora in 1815 and of Cosegüina in 1835, for instance, lasted less than a week.

Major outbursts on most strato-volcanoes are usually separated by long periods of repose that may last decades or centuries, during which they often give scant indication that they are still active. Montagne Pelée, for example, has been tranquil since 1932, and Pico Viejo in Tenerife since 1798. In other strato-volcanoes, however, persistent fumaroles indicate that hot magmas still linger close to the surface. Such is the state of Fossa on Vulcano, last active in 1890, of El Piton on Teide, which probably last erupted in 1492, and of Mount Rainier, whose last outburst occurred about 1300 years ago. Sometimes, too, solfataras form sulphurous crystals, like those collected from Popocatépetl by a lucky Spanish Conquistador, who was lowered in a basket into the fuming crater as the eruption in 1519 began to wane – and emerged unscathed. Some more considerate strato-volcanoes issue spluttering reminders that they are still alive in the form of sporadic, small-scale eruptions. Mount Hood, in the Cascade Range, has been quiet for over a century, but minor eruptions recorded in 1853, 1854, 1859 and 1865, and more powerful, intermittent activity probably between 1760 and 1810 – all show that the 40000-year-old volcano is far from extinct (Figs 10.1, 10.7).

The periods with little or no surface activity, should perhaps not be called

Figure 10.7 The summit of Mount Hood, with small domes on the left skyline and nuées ardentes deposits in the middle and foreground.

"repose" or "dormancy", because, in fact, they usually constitute vital interludes within the magma reservoir, where the ingredients of the next violent eruption are being prepared. The calm on the surface is often entirely deceptive. The behaviour of Vesuvius strikingly illustrates the significance of these repose periods. The longer the interlude of surface inactivity, the more explosive the magmas in the reservoir become, and the more violent is the subsequent eruption. The cataclysm of AD 79, for instance, brought many centuries of calm to a sudden and spectacular close, and the great, but less violent eruptions of AD 472 and 1631 followed at least 200 years of repose. The outburst of 1631 inaugurated a period of more persistent activity, most marked between 1872 and 1906, which was characterized by oft-repeated eruptions that were relatively mild *by Vesuvian standards*, during which quiet intervals never exceeded seven years. This period ended with the eruption in 1944. The subsequent calm interlude has been by far the longest since 1631 – indicating that the next outburst will be the most violent since that date.

Craters on strato-volcanoes

Active or dormant strato-volcanoes are usually crowned by craters, which are, however, smaller in proportion than those of cinder cones. But strato-volcano craters are more varied, chiefly because the explosions forming them have a much greater range in intensity. Erupting craters are probably funnel shaped, but impossible to survey! During repose many have the typical circular bowl shapes of the craters of Vesuvius, or Fuji, which is 700m across and 100m deep; others are wineglass-shaped like that on Popocatépetl, which is over 670m across and 200m deep.

The craters of higher strato-volcanoes, such as the Andean Nevados, may be hidden by ice, which finds them ideal hollows in which to nestle. Ice fills the summit crater on Mount Rainier, although hot fumaroles have melted its lower layers to form warm ice-caves. Sometimes, indeed, these exhalations are warm enough to stop any ice forming, and hence, even in Alaska, the crater of Katmaï is usually filled with warm waters. In wetter and warmer zones, craters readily collect rainwater and form substantial lakes, such as those on Galunggung and Kelut in Java. But, although lakes and ice-caps might enhance the beauty of many summits, they can also help form lethal lahars. The crater is usually the site of the most rapid changes on a strato-volcano. These changes can be both constructive or destructive – ranging from the development of explosion pits to the caldera formation that decapitated Tambora in 1815, Cosegüina in 1835, and Mount St Helens in 1980. Pico Viejo, for example, is crowned by a circular, ragged-rimmed crater some 750m across and 150m deep, whose floor is pitted by two large hollows, one 140m and the other 75m deep, blasted out by phreatic explosions. Holes on an altogether grander scale mark the summit cone of Etna. The Voragine ("the chasm") formed in 1945, and the Bocca Nuova ("the new mouth") initiated in 1968, have recently joined to form an abyss about 500m across, which, however, varies considerably in depth. They usually exhale no more than gas and steam, and can then be over 1000m deep. Sometimes, however, molten basalts surge almost to their very rims and form lava lakes, from which cinders and ash may explode. More rarely, as in November 1973, spectacular lava-fountains can spurt 2.5 km into the air, or phreatic explosions can be unleashed without warning, like those on 17 July 1960, and 12 September 1979.

Not all the events that take place within their summit craters are destructive, otherwise strato-volcanoes would cease to grow. Many destructive episodes are only negative intervals in a broad history of accumulation. Some – rare – strato-volcanoes usually only display relatively small-scale eruptions. The most persistent and most extraordinary of these is the crater of Stromboli, where basaltic eruptions occur from several vents at intervals of less than an hour. The crater of Pico perhaps provides a more typical example (Fig. 10.5). Although it is 500 m wide, it is now only about 30 m deep because it has been filled almost to the brim by pahoehoe lavas within the past millennium. The lavas form a distinct basaltic cornice on the crater rim that supports a small cinder cone, Piquinho, that rises about 60 m above it.

These crater-filling episodes are not limited to basaltic eruptions. At Cotopaxi, in the Andes, a low cone of andesitic fragments lies in the summit crater like a doughnut in a soup plate. The summit of South Sister in Oregon is marked by a small cone of basaltic andesite cinders and thin lava-flows, with a sharp-rimmed crater about 400 m across, which looks fresh enough to have been formed within the past thousand years. This cone already masks the crater from which the andesites and dacites forming the bulk of the cone were expelled. In Tenerife, El Piton, the cone crowning Teide, rises 150 m high, like a giant sandcastle inside – and almost completely buries – the former summit crater, La Rambleta (Fig. 10.2). El Piton itself encloses a funnel-shaped crater, La Caldereta, that is 70 m wide and 40 m deep. El Piton is a miniature strato-volcano in its own right, composed of layers of phonolitic fragments and lava-flows, often altered to brilliant white by fumaroles. Its last eruption, which spilled out glassy, black phonolitic lavas, was probably observed by Columbus on 24 August 1492 en route for the Americas. He saw "a great fire come out of the summit area of Tenerife", and calmed his apprehensive crew by explaining that it was "just like Etna".

Shifting central vents

During the long lives of many strato-volcanoes, their dominant central vents sometimes shift laterally by 1 km or more. As a result, the summit becomes a realm of varied relief, well illustrated by Shasta in the Cascade Range (Fig. 10.8). It rises to a height of 4316 m from a base at about 1000 m and is really composed of four overlapping cones of andesitic fragments and flows which together attain a volume of about 500 km^3 (Fig. 10.2). The first cone, now exposed on Sargent's Ridge, was formed over 200 000 years ago from a vent on the upper southern slopes of the present volcano. The second cone, which now remains as Misery Ridge, erupted 15 000–20 000 years ago from a vent 2 km to the north of its predecessor. A new vent then opened 3 km to the west, which constructed Shastina, the graceful cone rising

Figure 10.8 Shasta, California: Hotlum Summit on the left, Misery Ridge in the centre, and Shastina on the right.

to 3749 m that forms the southwestern summit of the volcano. Shastina erupted between 9700 and 9400 years ago and is composed of andesitic lava-flows, blanketed by nuée ardente deposits, with four dacitic domes in its 800 m-wide crater. Activity then transferred 4 km to the northeast and formed the Hotlum cone at the present summit. Most of Hotlum is less than 4500 years old, and is an assembly of stubby andesitic lava-flows 150 m high, many of which solidified on slopes of 20°. Hotlum also has a small crater containing a dacitic dome. Its latest eruption was probably that witnessed from the Pacific Ocean by the French explorer La Pérouse in 1786. It was almost certainly not the last.

Flank eruptions and planèzes

On many strato-volcanoes the activity of the central cones is supplemented, and at times superseded, by eruptions on their flanks. The distribution of these eruptions is often dominated by fissures, which usually radiate from the summit, or, as on Beerenberg and on Faial, for instance, follow directions determined by major crustal fractures (Fig. 6.17). The magmas can be either viscous or fluid. Flank eruptions can, for instance, expel evolved viscous magmas which have lost their volatiles after a major summit eruption. They then give rise to domes such as Montaña Rajada at the foot of Teide, or Colle Umberto on Vesuvius. Such eruptions are usually both small and infrequent. Fluid flank emissions are usually larger and more frequent.

They commonly rise from the lower parts of a reservoir, where fluid basaltic magmas are concentrated. Their vents may either branch from the main summit conduit, or they may extend independently from the reservoir – it is often difficult to distinguish between the two types. They are overwhelmingly basaltic in character, giving rise to cinder cones and long lava-flows. Etna has seen over 200 flank eruptions in the past few thousand years, whereas Klyuchevskoy and Pico can each boast about 150. As contemporary Etna demonstrates, flank eruptions can occur at the same time as summit activity, or after it has waned. Thus, for example, the basaltic eruptions on the periphery of both Cantal and Mont-Dore volcanoes took place after most of their central eruptions had ceased. They formed "planèzes" (the plains surrounding the mountains) where recent basalts intermingled with relics of basal plinths emitted millions of years before. The two phases of similar activity were thus often confused, and "planèzes" were usually considered to be ancient features until closer analysis revealed their youthful aspects. But not all strato-volcanoes develop a multitude of flank eruptions. Mount St Helens is bereft of satellites and Stromboli – or at least its emerged parts – has only six, including the islet of Strombolicchio (Fig. 5.8).

Flank displacements

The flanks of many strato-volcanoes have collapsed along major faults, which made straight step-like scarps facing outwards from the crest of the volcano. Both the northern and southern sides of Pico are let down along such faults, forming scarps of more than 1000m (Fig. 10.5); and the western promontory of Faial has foundered by over 500m (Fig. 6.17) The lower eastern slopes of Etna are also thrown down towards the Ionian Sea along a series of scarps, called "timpa". Less frequently, but no less spectacularly, faults sometimes let down sectors of a strato-volcano, like a key pressed down on a piano keyboard. The Sciara del Fuoco is the most famous example of such a sector, but Tenerife has two comparable features, forming the Orotrava and Güimar valleys, which lie back-to-back on opposite sides of the island.

These scarps, however, are by no means as large as the Valle del Bove, the great chasm on the eastern side of Etna that is 6km across and often 1000m deep. It is an enigma, whose origin remains in doubt after more than 200 years of study. It may have been caused by gigantic landslides, by faults, by repeated caldera-forming explosions, or by debris avalanches. It is one of those landforms that stimulates research by defying satisfactory explanation. Thus, strato-volcanoes are the most varied of volcanic forms and exhibit a great range of eruptive styles. Their complexity is such that they are, in many ways, the least understood of volcanic features on land. As a result, there is still no way at present of ensuring whether a strato-volcano is extinct, dormant, waning, or just waiting to unleash a cataclysm. This has grave repercussions for predicting their future behaviour.

11

Calderas

Calderas are often so serenely beautiful that it is hard to imagine that they have been produced by some of the most catastrophic events on Earth. Many develop within the space of a few hours. Huge volumes of material are erupted and huge volumes of material also sink into the crust. At least six of the ten caldera-forming eruptions in the past 10 000 years in the eastern Aleutian arc, for example, have each probably ejected more than 50 km^3 of ash and pumice. Fortunately calderas are not formed very often and, on average, only two or three are created every century.

A caldera is a large volcanic depression, at least 1 km in diameter and often much more, that is enclosed by nearly vertical walls facing onto a central flattish floor. Many calderas reach 25 km across and are bounded by walls 1000 m high; several are partly filled with lakes, such as Sete Cidades, Furnas and Fogo calderas in São Miguel; and some, like Santorini and Vulcano, have been invaded by the sea. Most are circular or elliptical in shape, but some, like , Azores and Aira, are more rectangular. Others, like the Caldera de las Cañadas in Tenerife and Monte Somma-Vesuvius, are asymmetrical because one sector of the walls is missing. It seems that those calderas which form most rapidly are defined the most clearly by steep walls. Those created in sporadic episodes have blunter, scalloped, or stepped edges. After a long recuperative pause, renewed eruptions often begin to fill many calderas, with landforms ranging, for example, from domes at Nysiros in Greece, to a cinder cone at Crater Lake in Oregon, and to two strato-volcanoes at the Caldera de las Cañadas. Fogo, in São Miguel, seems alone amongst European volcanoes, for example, in not having received subsequent eruptions on its floor – but its most recent caldera-forming episode probably only took place in 1563 (Fig. 11.1). Older calderas are hard to distinguish at all. Some, like Crater Lake caldera, have boundaries of outstanding clarity, but the outlines of many others, such as Povoação in São Miguel and Guilherme Moniz in Terceira, have been blurred by erosion; others, such as Cerro Gallan in Argentina, are so shallow and so large that they have only recently been identified through satellite images.

Five broad groups of calderas may be recognized, each with different origins. Four of the groups are dominated by collapse: on basaltic shields; on strato-volcanoes; on complex volcanic fields; or as a result of ashflows. The fifth, and most recently recognized, group is formed by debris avalanches and, alone in the genre, *always* displays notably asymmetrical outlines.

188

Figure 11.1 Fogo caldera, São Miguel.

Basaltic shield calderas

Basaltic shield calderas are typified by the Hawaiian volcanoes, but they also occur in similar oceanic shield environments such as Réunion, Iceland and the Galápagos Islands (Fig. 5.7). They are exceptional amongst calderas because they are not usually associated with violent eruptions, although Kilauea caldera is believed to have developed in association with an uncommonly violent explosion (for Hawaii) in about 1790. Collapse occurs at the summit when a piston-like mass subsides down nearly vertical faults. Foundering is often brought about at Kilauea when magma is syphoned away from the main reservoir and erupts in subsidiary rift zones. The Hawaiian calderas are shallow and are quite small compared with the great bulk of their parent volcanoes. Kilauea caldera is only 150m deep, whereas Mokuawe-oweo caldera on nearby Mauna Loa is 180m deep, and both are just under 5km across. Within the broad calderas on such shields, much smaller pits often develop, such as the Bory and Dolomieu craters in the Enclos Fouqué caldera in Réunion, in which lavas from time to time form molten lakes. Lavas also erupt from concentric and radial fissures and supply the vast majority of the frequent eruptions of Piton de la Fournaise. Created about 4745 years ago, the Enclos Fouqué is the latest of three successive calderas formed on its summit.

Strato-volcano collapse calderas

Strato-volcano collapse calderas are by far the most numerous of those yet identi-fied, perhaps because they are formed on the summits of big mountains and are thus easiest to see. Such calderas have often been created several times during the lifetime of many typical strato-volcanoes, but good observations of exactly how they form are rare, if only because anyone witnessing the event closely enough would not live to record much information. Most are between 2m and 13km across, and they can range in depth from 100m to over 1000m. They are typically linked to andesitic, dacitic or rhyolitic eruptions, especially of pumice, which occur when the summit area founders into the emptied magma reservoir (Fig. 11.2). It is hard to tell whether the eruption causes the collapse or vice versa. Not that these eruptions themselves should be underestimated. In April 1815, Tambora not only formed a caldera over 6km in diameter, but also expelled 125km^3 of fragments.

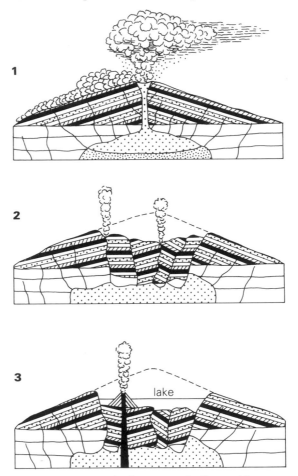

Figure 11.2 The formation of collapse calderas.

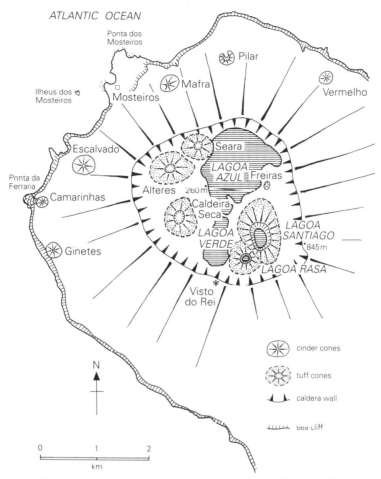

Figure 11.3 Map of Sete Cidades volcano and caldera, São Miguel.

Some fine examples of strato volcano collapse calderas are found in the Azores. Although none was formed during the 500 years of historic time, several have retained remarkably sharp outlines, with nearly vertical walls, on the crests of decapitated mountains. The calderas of Graciosa and Faial, of Santa Barbara in Terceira, and of Sete Cidades, Fogo and Furnas in São Miguel, are so fresh-looking that they cannot be more than a few millennia in age. The view from the Visto do Rei on the rim of the Sete Cidades caldera in São Miguel is the most photographed in the Azores. It shows an almost circular caldera, some 5 km in diameter, with steep walls, whose lower slopes are often clothed in cryptomeria, hydrangea and hedychium. Its floor is 251 m above sea level and the encircling walls rise between 250 m and 600 m higher. It is occupied by several cones and two adjoining lakes, the Lagoa Azul and the Lagoa Verde (Fig. 11.3). The glamour of this, the deepest caldera in the Azores, is enhanced for some by the harmless legend that seven cities – "sete cidades" – were

191

engulfed in its formation. The strato-volcano is the most recent in São Miguel, probably beginning about 210000 years ago, and it was chiefly constructed of layers of first andesitic and then trachytic lava-flows. The caldera has a volume of 6km³. It collapsed 22000 years ago when a mass of trachytic pumice erupted in Plinian columns and nuées ardentes and reached a thickness of 60m towards the southeast. Eruptions resumed along a ring fracture about 1km inside the caldera walls, and they took place, at intervals, for about 17000 years. They formed the tuff rings of Seara, Caldeira do Alferes and Caldeira Seca, and the tuff cones containing the lakes of Lagoa Rasa and Lagoa Santiago, as well as the larger craters now occupied by the waters of the Lagoa Azul and the Lagoa Verde. These seven features could, in fact, lie at the origin of the Seven Cities legend. The Rasa and Santiago vents have been the source of all but the first and the last of the dozen eruptions, often of Plinian proportions, within the caldera in the past 5000 years. The first eruption originated from the Seara vent, and the last, which has been dated to AD 1287, probably formed the tuff ring of the Caldeira Seca. Its weak tuffs were vigorously gullied even before the rapidly growing vegetation of the Azores eventually got a hold, and Caldeira Seca thus looks older than its years. The most recent of all the eruptions in the Sete Cidades caldera built Freiras, the cinder cone 75m high alongside the Lagoa Azul in 1444. In spite of all the commotion in the past 5000 years, however, only the weakest fumarole activity now occurs within the caldera, but if past trends are maintained, another eruption could occur there in less than 200 years.

The island of Nisyros in the Aegean Sea is a strato-volcano with a central caldera (Fig. 11.4). The strato-volcano was built up above the level of the Aegean Sea about 66000 years ago, and it had probably risen to a height of 1000m about 24000 years ago. Its crest then collapsed in conjunction with a massive Plinian eruption of dacitic pumice, which still commonly forms a blanket 100m thick on the upper slopes of the island. Nisyros probably lost its upper 500m in consequence. The caldera floor lies only 100m above sea level, and it is surrounded entirely by steep walls ranging from 150m to 400m in height, where blue-grey andesitic and dacitic lava-flows stand out in vertical cornices. Its original diameter reached 3km across, but, after a period of repose, three large and two smaller bristling domes were extruded which now occupy half the original floor area. The largest, the St Elias dome, is almost 600m high and ranks amongst the bulkiest in Europe. To judge by the freshness of their forms and their little-weathered surfaces, these domes can scarcely be more than a few thousand years old. The latest eruptions took place in 1873, building tuff cones, 100m high, at the foot of the domes, while a phreatic explosion blasted out a circular hollow, 300m across and 25m deep, in the caldera floor nearby. This pit, Stephanos ("the crown") provides the sole contemporary activity in the shape of fumaroles, mudpots and hot springs. This modest hydrothermal spectacle is called "the volcano" in the explanations given to the tourists venturing forth on exotic excursions from Kos.

Figure 11.4   Nisyros caldera, Greece, with the St Elias dome on the left and Stephanos explosion crater in the foreground.

Volcanic field calderas

The calderas formed in volcanic fields are commonly larger than those developed on strato-volcanoes and often exceed 10km in diameter. These calderas are created, however, in much the same way as those on strato-volcanoes, except that collapse is often accompanied by even more violent eruptions of ash and pumice. A further difference springs from the nature of the volcanic area before the eruption started. A strato-volcano would have risen to a conical crest, whereas the volcanic field would be composed of several neighbouring, but often independent, volcanoes. Thus, after the eruption, a caldera formed in a volcanic field often has a more varied, scalloped, outline, and a more undulating rim where foundering transected different volcanoes. Re-appraisal of Santorini, for example, has revealed that the Minoan caldera partly engulfed a number of lava shields and small strato-volcanoes in the northern half of the island (Fig. 6.15).

The great eruption of Krakatau similarly affected a volcanic field rather than a single strato-volcano. Before the eruption began on 20 May 1883, Krakatau consisted of three adjoining cones. Rakata, the largest and oldest, formed a basaltic cone in the south, with the andesitic cone of Danan in the centre and the younger andesitic cone of Perbuwatan in the north. Perbuwatan started the eruption; Danan joined in a month later; and Rakata followed on 11 August. The cataclysm began at 13.00 on 26 August 1883, and continued throughout two days of din and darkness,

193

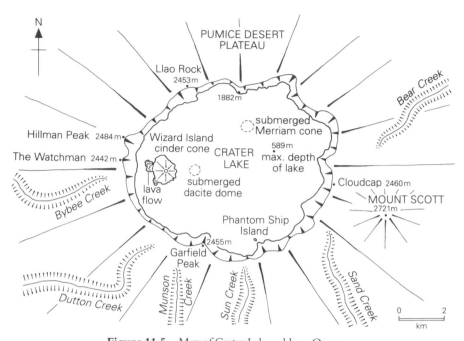

Figure 11.5 Map of Crater Lake caldera, Oregon.

with four vast eruptions of pumice, culminating at 10.02 on 27 August in an explosion that made probably the loudest noise heard in the world since the Minoan
eruption of Santorini, and which still sounded like gunfire 4925 km away on Rod-
rigues Island in the Indian Ocean. The blasts broke windows in western Java and
disturbed the pressure gauge, 150 km away, in the gas works in Batavia (Jakarta).
Each explosion, with nuées ardentes and debris avalanches, accompanied the collapse and engulfment of the caldera into the Sunda Straits, and together they generated the most devastating tsunamis of modern times. At length, when the eruption
ceased, Rakata had been sliced in two when the caldera collapsed, and only its
southern half remained. Danan and Perbuwatan had vanished entirely. The flooded
caldera was over 6 km across and more than 18 km³ of ash and pumice had been
expelled. The ash ejected into the stratosphere circulated around the world, giving
the eruption an epilogue of brilliant sunsets for several years.

An eruption of similar proportions formed Crater Lake caldera from a volcanic
field in southern Oregon (Fig. 11.5). Like Santorini, it developed in the midst of a
cluster of adjoining andesitic strato-volcanoes, domes and shields, which began
erupting about 420 000 years ago. This complex, long thought to have been a single
volcano, is called Mount Mazama, but different parts of the field erupted at different
times. Mount Scott in the south is 420 000 years old, Llao Peak and Garfield Peak
are both over 110 000 years old, and Hillman Peak in the north is about 70 000 years
old. The climactic outburst came 6845 years ago. An enormous Plinian column and
nuées ardentes spread rhyodacitic pumice and ash all around the vents. The nearby
valleys, such as Annie Creek, were choked with over 80 m of pumice for at least

194

10km, and, in all, more than 50km³ of materials were expelled. Fine ash was carried over most of the western parts of North America. The caldera foundered by some 1250m and sliced the peaks of Llao, Hillman and Garfield in two (Fig. 11.6). Subsequent eruptions have emitted andesitic lavas about 300m thick, a cinder cone over 400m high, and a small rhyodacitic dome – all of which remain submerged – as well as Wizard cinder cone, formed about 6000 years ago, which rises 250m above the lake. The lake, now 588m deep, occupies the elliptical caldera, measuring 10km by 8km, which is enclosed by almost vertical walls rising 600m above the water. The serenity and pure, startling sapphire blue of Crater Lake belie its violent origins, and it is now the fitting centrepiece of one of the finest national parks in the United States.

Figure 11.6 Crater Lake caldera, Oregon, with Hillman Peak on the left, Llao Peak on the right, and Wizard cinder cone in the centre.

Debris avalanche calderas

Strato-volcanoes are also the main locations of debris avalanche calderas, which have only been fully recognized since the eruptions of Bezymianny and Mount St Helens. Debris avalanches are now also thought to have played an important rôle in the eruption of Krakatau and in the formation of the asymmetrical Caldera de las Cañadas in Tenerife. A debris avalanche caldera is the hole left in the volcano when a debris avalanche has departed from its side in one violent eruption often lasting only a few minutes. They differ from collapse calderas in their size, deposits and shape (Fig. 6.7). They are smaller than collapse calderas and those so far identified range from 1.5 km to 15 km across. The debris avalanche is also not engulfed, but is deposited in a fan-shaped hummocky layer at the foot of the destroyed flank of the volcano. The volume of the caldera should equal that of the debris avalanche blasted from the volcano. The debris avalanche caldera is also bordered by steep, usually vertical walls. Collapse calderas may sometimes be asymmetrical, whereas debris avalanche calderas are always asymmetrical and odeon, or horse-shoe, shaped. One sector gapes open where the walls have been blown away down to the very floor of the caldera (Fig. 6.6). This gap then dictates the path taken by any subsequent eruptions of nuées ardentes or lava-flows.

Ashflow calderas

Ashflow calderas are usually more than 10 km and sometimes over 30 km in diameter, but they are often relatively shallow and most of them were formed on extensive ashflow plateaux rather than on volcanic cones. They are generally the world's largest calderas and are most often associated with rhyolitic eruptions. The volume of materials erupted can be enormous. Some 3000 km^3 of ashflows were emitted from the calderas at Yellowstone Park and at Taupo in New Zealand, but, like the Yellowstone caldera, they often lack clearly defined outlines – which is why satellite photography has been so useful in their detection. All these calderas collapsed along with, or very soon after, an enormous ashflow eruption that left the crust unsupported above the magma reservoir. About 3430 years ago, a vast ashflow eruption created the caldera of Aniakchak, which is the most spectacular in the Alaska Peninsula, some 10 km across and 1000 m deep. This is one of the eruptions that could have been responsible for the acidity concentrations in the cores from the Greenland ice-cap attributed to the Minoan eruption of Santorini. An ashflow eruption also appears to be the origin of Aira caldera in southern Kyushu in Japan. It collapsed about 21 000 years ago into a large fault-bound rectangular hollow, 25 km across, in the midst of a vast plateau of freshly erupted ashflows. The caldera is now flooded by Kagoshima Bay where its floor is more than 100 m deep. It was on its southern rim that Sakurajima first began erupting as a volcanic island about 13 000 years ago.

Perhaps the most scintillating of all calderas is occupied by Lake Atitlan in Guatemala. Lake Atitlan probably deserves its reputation as one of the most beautiful in the world – although admirers of Crater Lake could propose a serious competitor. Not least amongst the reasons for its beauty is the way in which it grew. In the past 14 million years, three similar calderas have collapsed, each slightly to the south of its predecessor, and Lake Atitlan lies in the third. About 85 000 years ago, it foundered some 1000 m into the crust accompanied by the eruption of the Los Chocoyos rhyolite pumice, which covers vast areas of Guatemala and has a volume of some 270 km^3. The lake then formed in the caldera has been reduced to half its initial area of about 260 km^2 because three andesitic strato-volcanoes subsequently erupted and now rise over 2000 m above its southern shores: San Pedro and the twin cones of Toliman and Atitlan (Fig. 10.4). Atitlan is certainly still active, for it erupted in 1469

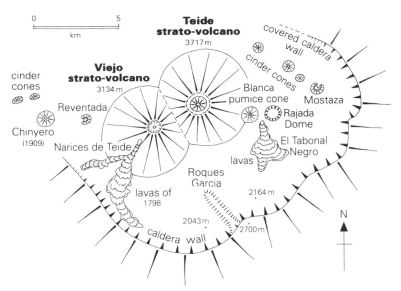

Figure 11.7 Selected types of volcanic filling in the Caldera de las Cañadas, Tenerife.

and sporadically again between 1826 and 1856, and its companions are most probably only dormant.

Caldera filling

As Lake Atitlan shows, the formation of a caldera does not usually end volcanic action. More eruptions indeed are only to be expected in zones riddled with faults situated just above a shallow, if temporarily emptied, magma reservoir. After a period of repose that may last for decades or centuries while magma again gathers in the reservoir below, eruptions then resume. They occur either within the caldera, or around its highly fractured edges and are usually less explosive than the eruptions creating the caldera. Thus, cinder cones, such as Wizard Island in Crater Lake and Anak Krakatau – or a mixture of domes and cinder cones such as the Kameni Islands inside the caldera of Santorini – often develop. Volcanic filling can start rapidly: a dome began to grow within the caldera of Mount St Helens within six months of the eruption in 1980, and Etna has filled both the Ellittico caldera and the Caldera del Piano within the past 10000 years. Sakurajima is fast filling Aira caldera and it is now one of the most flourishing volcanoes in the world, having produced more than 5000 eruptions since 1955.

The Caldera de las Cañadas in Tenerife demonstrates how varied this filling can be (Fig. 11.7). It foundered less than 170000 years ago at the summit of an enormous strato-volcano occupying much of central Tenerife, and stretches 17km from east to west and at least 12km from north to south. The clear atmosphere and sparse vegetation bring out the varied colours of the different rocks as if they had been painted. Steel blue phonolites streak down Teide, pale yellow pumice piles and dark red cinder cones decorate its base, and ragged brown phonolitic and glassy black basaltic flows spread across the caldera floor. The southern rim of the Caldera de las Cañadas is an impressive, nearly vertical cliff, 20km long and 700m high, that curves like a great wave about to break over the caldera, but its northern rim, if, indeed, it exists, is covered by the twin strato-volcanoes of Pico Viejo and Teide. Both are substantial cones in their own right, for Teide rises 1715m and Viejo 1134m above the caldera floor and together they have a total volume of 150km³. Both are still active and fumaroles still issue from the summit of Teide. Teide is attended by seven satellite volcanoes, five of which lie on a circular fissure around it. These satellite vents produced phonolite domes, such as Montaña Majua and Montaña Rajada, which also gave off a flow of phonolitic obsidian; some vents produced only blocky phonolite flows, such as El Tabonal Negro, which extends in a lobe 3km across the caldera floor; and Montaña Blanca is a cone of yellow pumice fragments, 300m high, from which blocky phonolitic lava-flows have also been emitted (Fig. 11.8). Many cinder cones and basaltic lava-flows have also been developed on the fringes of the caldera, often along fissures tapping deep-seated magmas independent of the Teide reservoir. Chinyero was the site of the most recent eruption in Tenerife on the western fringe of the caldera in November 1909. It formed a cone 50m high and a basaltic lava-flow some 4km long. The last eruption truly within the caldera took place

Figure 11.8   Caldera de las Cañadas, Tenerife. The dome of Montaña Rajada is in the foreground, and the pumice cone of Montaña Blanca is in the middle ground, with a lava-flow emerging from its base on the left. The strato-volcano Teide, crowned by El Piton, rises in the background.

from 9 June to 8 September 1798 at the Narices ("nostrils") de Teide, (situated, in fact, on the slopes of Pico Viejo). It contributed a long basaltic lava-flow, stretching in a slabby expanse to the caldera floor where its snout congealed in a small lava lake.

Calderas are not only filled by repeated volcanic eruptions, for their floors can also be raised by resurgence. Resurgence happens when the centre of a large caldera is arched up by a renewed upsurge of magma from the depths. Cerro Redondo in the Valles caldera in New Mexico and Samosir island in the Toba caldera in Sumatra are resurgent central domes. The arching may also crack the crust and much smaller domes of viscous lava can then be extruded onto the surface. In the Long Valley caldera in eastern California, which collapsed 720000 years ago, resurgent uplift of about 5 cm per year is being measured at present. As the magma reservoir lies only 6 km below the surface, a new eruption cannot be ruled out – although, at present rates, it might not occur for several thousand years.

The formation of the caldera occupied by Lake Toba in Sumatra deserves special mention as an example of an extraordinary eruption which dwarfs most others. This is one of the largest calderas on Earth, 100 km long by 30 km wide, and a grandiose eruption of 2000 km^3 of rhyolitic ash was associated with its creation in the space of perhaps two weeks, about 73500 years ago. The ash blanketed not only 25000 km^2 on land, but also covered 5 million km^2 of the Indian Ocean floor as far as India. The eruptive column may have reached a height of over 45 km. So much fine material was discharged into the stratosphere that it may have accelerated the onset of the last glacial period in the Ice Age by perhaps lowering temperatures by some 3°C or 5°C in the northern hemisphere for about a year, and maybe even decreasing summer temperatures in northern Canada by as much as 10°C or 15°C, thereby helping great volumes of ice to accumulate there.

Eruptions of similar size had previously occurred in North Island, New Zealand and in a few American centres several million years before. Luckily no comparable eruption has happened since then because its effects on humanity, not to mention the natural world, would be incalculable.

199

12

Hydrovolcanic landforms

Hydrovolcanic landforms develop when an external source of water interacts with magma approaching the Earth's surface. The large areas covered by water, ice, snow, and frozen ground, abundant precipitation, and subsurface aquifers, as well as innumerable paths provided by fractures in the Earth's crust, all combine to ensure that some interaction between water and magma during most volcanic eruptions is almost inevitable at present. It was even more likely to have occurred during the ice ages in the past 2 million years when ice-sheets and glaciers expanded and wetter pluvial periods in semi-arid areas often led to the formation of large lakes, notably in the American West.

It is the depth, amount and nature of the waters meeting the rising magma which largely determine the ensuing type of hydrovolcanic activity and the landforms created. Water and magma interact differently in submarine and subglacial, phreatomagmatic, Surtseyan and hydrothermal activity. At one end of the spectrum, eruptions in deep oceans are broadly limited to emissions of pillow lavas and mild gaseous emanations because explosions are prevented by the hydrostatic pressure exerted by the deep water. In this case the influence of the water is prolonged and non-explosive. At the other end of the spectrum, rainwater or snowmelt, penetrating fissures, may become superheated on contact with the magma or magma-heated rocks. The sudden transformation of water into steam in a limited space generates a violent explosion that clears a conduit and could then precipitate a magma eruption. In this case the influence of the water is ephemeral, strongly localized and notably explosive. Hydrothermal activity illustrates a third situation, which can often last for centuries, but the water only makes contact with heated rocks and thus violent explosions of steam are rare.

Where water is shallow, or in limited or intermittent supply on land, the interaction between water and rising magma provokes the greatest and most rapid changes. These conditions generate phreatic or phreatomagmatic and Surtseyan eruptions, which create hydrovolcanic landforms in the strictest sense – and form explosion craters (or maars), tuff rings, and tuff cones. Most hydrovolcanic landforms develop from basaltic magmas, which, without water , would produce Strombolian eruptions. Comparable features are also produced by brief eruptions of more evolved magmas, such as trachytes or dacites, but the "typical" hydrovolcanic characteristics in prolonged eruptions of evolved magmas are usually subsumed in the even more violent outbursts precipitated by their exploding magmatic volatiles. Nevertheless, hydrovolcanic influences have been revealed, for example, in the Minoan eruption of Santorini and at the Fossa cone of Vulcano. Hydrovolcanic activity therefore achieves its specificity where water and basaltic magmas interact, adding enough violence to the eruption to generate distinctive landforms.

The chief differences amongst hydrovolcanic landforms caused by basaltic eruptions are determined by the depth of the explosive contact between water and magma, which, in turn, is largely influenced by the duration of the water supply and the relative volumes of water and magma. Because these variables themselves have no hard and fast limits, there are no firm distinctions between the resulting landforms – and they have almost as many different definitions as scientists who have studied them. Maars seem to form always on land; tuff rings and tuff cones are erupted in, or very close to, shallow water, or above aquifers. Phreatic and phreato-magmatic eruptions produce maars and some tuff rings and they cease when the (usually limited) water supply is cut off. More prolonged Surtseyan eruptions generate tuff cones and most tuff rings and are often related to large supplies of standing water. Thus, they only usually stop when the magma stops rising, or when so much material has been erupted that water can no longer reach the vent. The theme pervading throughout these variables is the environmental control of their distribution exercised primarily through the type of water supply to the vent.

Maars

In phreatic or phreatomagmatic eruptions, the country rocks are shattered and founder into a trumpet-shaped conduit forming a diatreme piercing the crust (Fig. 12.1). A deep, rain-filled circular crater, or maar, is formed on the land surface that may be 200m deep and 1km across and bounded by a steep, inward-facing wall, exposing the country rock (Fig. 12.2). This wall is often surmounted by a ring, or crescent, commonly reaching 10–30m in height, which is made up of layers composed very largely of fragments of country rock, sloping at about 10°–15° away from the crater. There is a sharp contrast between the steep wall facing onto the crater and the gentle slopes extending outwards from its crest. The initial beds of shattered country rock are coarse because they are expelled when the main, throat-clearing explosion blows the conduit through the substratum. Subsequent beds are thinner and composed of finer fragments which may include a little new lava. Although maar-forming explosions are violent, the volume of fragments ejected is usually much less than the volume of the craters formed. Thus, maars seem to be chiefly caused when the conduit walls collapse and form a diatreme. Old diatremes exposed by erosion have similar trumpet-shaped conduits, 500m or more across, which extend deep into the crust and are choked with a mixture of volcanic and non-volcanic fragments commonly ranging from 1mm to 1m across.

The relationship between diatremes and maars can be deduced from the Plateau de Gergovie in Auvergne, where basaltic eruptions took place through sodden marls

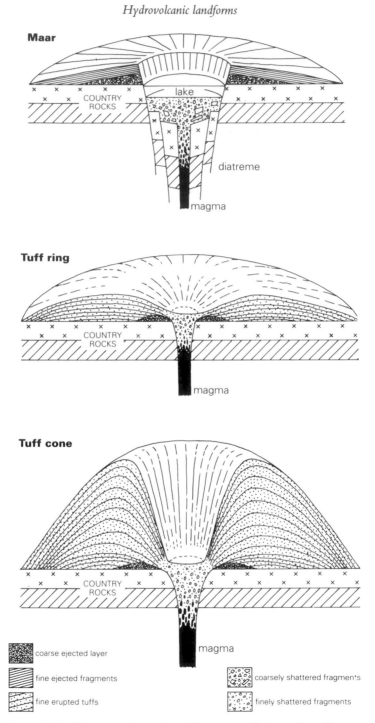

Figure 12.1 Diagrammatic sections of a maar, a tuff ring, and a tuff cone.

Figure 12.2   A maar near Reykjavík, Iceland (courtesy Margot Watt).

beneath a shallow lake. The diatreme, exposed on the eastern flanks of the plateau, is choked with shattered, churned up, fragments of brown basalt and buff marl, forming a peppery looking mixture known as "pépérite", whose true origin has now apparently been established after causing argument since the infancy of geology. Two successive, intersecting maars were formed – the first 19 million years ago, the second 16 million years ago. They would have been long eroded away, but they were filled by lavas from neighbouring vents (forming secondary lava lakes), which preserved their outlines and now cap the plateau. Maars are usually soon eroded away or filled with sediments. Old maars are thus hard to detect, and, although many must have been formed throughout geological time, all those that are most obvious at present were created relatively recently, often less than 100 000 years ago. For example, the 156 000-year-old maar which has the honour of underlying the main square in Clermont-Ferrand (in Auvergne) has only latterly been identified because it has lost its typical forms through infilling and erosion.

The Gour de Tazenat is the finest maar in the nearby Chain of Puys. It was formed when waters from the Rochegude stream filtered down an old fissure to meet rising basaltic magma. The ensuing explosions created an almost circular crater, 900 m across, now filled by a lake some 66 m deep. It is surrounded by a steep wall of country rock, topped by a thin rim of beds of angular fragments dipping outwards at angles between 5° and 10°. Beds largely composed of glassy, dark basaltic fragments alternate with those of pale, shattered country rock. The remarkable sharpness of the Gour de Tazenat belies its age, for, if its recent dating to 90 000 years ago is correct, it was formed by one of the oldest eruptions in the Chain of Puys.

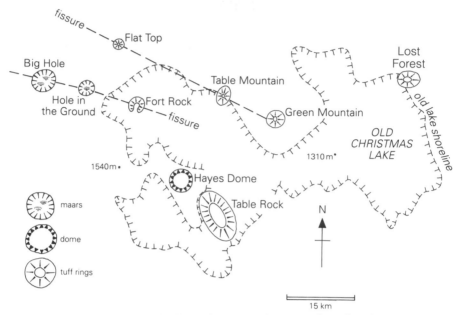

Figure 12.3 Selected volcanic features at Christmas Lake Valley, Oregon.

In southwestern Oregon, phreatomagmatic basaltic eruptions created two large explosion craters on the same fissure northwest of Christmas Lake Valley (Fig. 12.3). Big Hole, formed 20 000 years ago, and Hole in the Ground, formed between 13 000 and 18 000 years ago, were generated when water infiltrated to hot basalts during a pluvial period during the last ice age. Hole in the Ground was formed by four successive blasts, probably within a single day. The crater is 1.5 km across and its floor now lies more than 112 m below the surrounding country. The rim surmounting the almost vertical crater wall is between 35 m and 65 m thick and 90 per cent of the fragments are composed of country rock, some in masses 8 m across, although most are finely shattered. The upper parts of the conduit collapsed with each explosion; each collapse blocked the vent; and each succeeding explosion cleared it again until no more water was left. The craters lack lakes only because the area is now so dry.

Most maars are formed from basalts, but brief eruptions of more evolved magmas can also take part in their development. The dozen Inyo craters in eastern California were formed by predominantly phreatic explosions about 550 years ago. Less than 5 per cent of the ejected fragments came from the rising rhyolitic magma and the remainder from the country rocks. The eruptions had a violent start and ejected large blocks, but they waned quickly until only fine ash derived from a small amount of new magma was expelled. South Crater is the largest hollow, 250 m across and 55 m deep, and, unusually, its volume is approximately equal to that of the fragments ejected, which indicates that it was chiefly formed by explosion with little collapse.

Tuff rings

Tuff rings are best formed by Surtseyan eruptions generated when magma rises beneath a lake or shallow sea, or in areas of abundant groundwater (Fig. 12.1). Explosive expansion of the steam is shallower, less confined, and thus weaker, but more prolonged, than in phreatomagmatic eruptions because more water is available for a longer period. More magma also reaches the surface so that its shattered fragments thus constitute the bulk of the tuff rings.

If the circular crater hollow is the most striking characteristic of maars, attention focuses in tuff rings on the smooth, low mound of fragments usually enclosing a wide, shallow crater that often stretches from 500m to 1km across (Fig. 12.4). The crater, however, rarely contains a lake, not only because its floor is higher than the surrounding country, but also because the encompassing fragments are often permeable. Tuff rings often have the approximate proportions of a tyre or doughnut and in many cases both the inward-facing and outward slopes are equally gentle, and rarely exceed 12° in freshly erupted examples. Their crater-ward slopes, however, have sometimes slumped under gravity so that they are nearly vertical.

Tuff rings commonly rise between 40m and 60m in height but may reach 150m if the eruption lasted an unusually long time. They are composed of tuff layers ranging from 1 to 5cm in thickness which generally dip at angles from 1° to 10°, sloping inwards towards the crater and radiating away from the tuff ring on its outer flanks. They are laid down hot and dry at temperatures exceeding 100°C, by surges of superheated steam and usually remain in unconsolidated layers. The first, most

Figure 12.4   Tuff ring near Açigöl, central Turkey.

205

violent explosions eject thin layers of coarse fragments, but subsequent weaker explosions form beds of fine, angular, dark-coloured, glassy, basaltic fragments. Fresh magma accounts for at least 80 per cent and often over 90 per cent of the fragments, with the remainder composed of shattered country rock. The eruptions continue for several weeks or a few months until the magma no longer reaches the surface.

Christmas Lake Valley represents the floor of an old lake, but its landscape is dominated by basaltic tuff rings, which include Flat Top, Lost Forest, Green Mountain, Table Rock, Table Mountain and Fort Rock (Fig. 12.3). Fort Rock was built between 18 000 and 13 600 years ago when basaltic magma exploded through the lake and was shattered into fine orange-brown glassy, lapilli-sized tuffs in beds from 1 cm to 1 m thick that constructed a tuff ring 60 m high and 1 km across (Fig. 12.5). Its crater floor stands 10 m above the old lake bottom and its inward-facing walls have been much steepened by slumping. Fort Rock really does resemble a medieval fortress because the waves of the old lake eroded its outer slopes and breached its southern sector. Thus, slumping and erosion have made its originally gentle inner and outer slopes untypically steep. Table Rock, 30 km southwest of Fort Rock, is made up of a group of two large and six small tuff rings covering an area 9 km long and 6 km wide (Fig. 12.3). The explosions varied greatly in intensity so that the ochre beds ranging from ash to lapilli-sized tuffs make up beds from 1 m to 1 mm in thickness. Like Fort Rock, Table Rock was eroded by the waves of the ancient lake so that its outer slopes are now steeply cliffed.

Figure 12.5   Fort Rock, Oregon.

Tuff cones

Surtseyan eruptions construct tuff cones when continuous supplies of large quantities of water enter into contact with rising magma (Fig. 12.1). Their formation requires a more abundant water supply, and explosions at greater depths than those needed to build tuff rings, but both features develop best in lakes and shallow seas. Tuff cones are made up of tuffs which are dense, well rounded and commonly about

206

2.5 cm in diameter. They form beds ranging in thickness from 3 cm to 1 m according to the duration and intensity of the explosive spasm that ejected them. They emerge at temperatures below 100°C so that the steam entrapped is no longer superheated. Water therefore condenses on the fragments, which are thus often wet when they come to rest, whereupon physical and chemical changes can then weld them into a stiff, consolidated, tan-coloured mass. Most fragments are composed of new magma and only between 5 per cent and 10 per cent are derived from the country rocks.

The eruptions may last for several months and soon build up large, rather squat tuff cones, often rising from 100 m to 300 m in height and stretching from 1 km to 1.5 km across. Their smooth outer slopes, like the dips of their component beds, frequently average about 25°. Tuff cones have deep, broad craters, narrower and steeper than those of tuff rings, but deeper and wider than those of unbreached cinder cones. The crater walls may be as steep as 40°, and are often notably funnel shaped, with a steeper-sided lower section, some 250 m across, splaying out to an upper rim about 750 m in diameter. Their crater floors commonly lie less than 100 m, if at all, above the surrounding area. They often resemble craters formed by Vulcanian eruptions, where water interference has also been recently indicated.

The sea has breached the craters of many tuff cones. Open sectors develop as soon as the cone starts to grow and they can continually let water into the vents and thus ensure that Surtseyan eruptions can persist. Of all hydrovolcanic features, tuff cones have the closest affinities with Strombolian cinder cones, not only in their dominant shapes, but also in the duration of their eruptions. Most tuff cones are formed from basalts, but they sometimes develop from more evolved magmas. Within the Sete Cidades caldera of São Miguel, for example, trachytic magmas have built the tuff cones enclosing the Lagoa Rasa and the Lagoa Santiago (Fig. 11.3).

The shores of many volcanic islands are marked by tuff cones, one of the finest of which is Monte Brasil in the Azores, which shelters the capital of Terceira, Angra do Heroismo (Fig. 12.6). It was formed by a basaltic eruption in waters 50 m deep, which built up layers of fine, ochre or brown tuffs that make a cone reaching 205 m above sea level. Its broad crater is 170 m deep and has remained open to the south, but, although it is extensively cliffed on its seaward sides, Monte Brasil still has a diameter of more than 1 km after over 500 years of marine attack.

Hydrovolcanic eruptions and changing environments

Hydrovolcanic explosions and their landforms are subject to strong environmental controls, which may vary from time to time in one place, or from place to place where a single fissure extends from wet to dry areas. In any given eruption, water might be prevented from reaching the vent by a copious upsurge of magma, by the construction of an impermeable tuff cone, or, indeed, if the water supply were to be removed by, say, natural lake drainage. In each case, provided magma continues to rise, a basaltic eruption would then assume its more common Strombolian activity. Thus, in a single vent, one upsurge of magma could give rise to successive types of

Figure 12.6   Monte Brasil tuff cone, Angra do Heroismo, Terceira.

eruption and to superimposed landforms. The Rothenberg cones in the West Eifel of Germany show three alternating episodes of phreatomagmatic and Strombolian activity. The week-long eruption of Monte Nuovo in September 1538 in the Phlegraean Fields west of Naples began with Surtseyan emissions when water infiltrated the vent from nearby Lake Averno (Fig. 6.2). A tuff cone formed in two days, which cut off the water supply to the vent, so that later in the week Strombolian eruptions covered the tuff cone with cinders. The dimensions of Monte Nuovo, however, reflect the predominance of Surtseyan activity, for the cone is 133 m high and 800 m in diameter, with a crater 400 m wide and 120 m deep.

A more complex sequence occurred at Puy de Pariou in the Chain of Puys (Fig. 12.7). The first eruptions about 9000 years ago formed a basaltic cinder cone which now only survives at the southwestern base of the volcano. The second eruptions soon afterwards, allied to a water supply of unknown provenance, constructed a tuff ring 750 m wide and about 50 m high. The repercussions of the third eruptive episode, about 8240 years ago, were more complicated, because the removal of the water supply enabled both cinders and lavas to be erupted simultaneously. The lavas accumulated in a lava lake, about 50 m deep, inside the tuff ring, but they surged up so strongly that they pushed aside its northern sector and left it stranded 300 m away as a distinct mass, called Puy de Petit Fraisse, which was long thought to be a separate cone (Fig. 9.2). At the same time, in the southern part of the tuff ring, other Strombolian eruptions were building New Pariou, a cinder cone 160 m high, with a sharp-rimmed crater, 93 m deep and 300 m in diameter.

Just as the relationships between rising magma and water can change with time

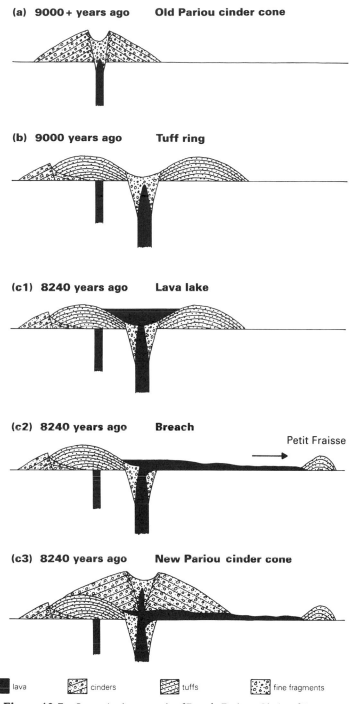

(a) 9000 + years ago   **Old Pariou cinder cone**

(b) 9000 years ago   **Tuff ring**

(c1) 8240 years ago   **Lava lake**

(c2) 8240 years ago   **Breach**

Petit Fraisse

(c3) 8240 years ago   **New Pariou cinder cone**

lava   cinders   tuffs   fine fragments

Figure 12.7   Stages in the growth of Puy de Pariou, Chain of Puys.

in any given spot, so they can alter where similar magmas erupt along fissures extending from wet into dry environments. Thus, in Christmas Lake Valley, the fissures that produced the Table Rock tuff rings within the lake formed the low dome of Hayes Butte above its former shores (Fig. 12.3).

One of the clearest examples of such a fissure sequence occurs at the southern end of the Chain of Puys (Fig. 12.8). About 10 700 years ago, waters infiltrating from a stream into a fissure caused a phreatomagmatic basaltic explosion that created a maar, about 120 m in diameter, which was later filled to form the Narse ("marsh") d'Ampoix. A similar basaltic eruption in a dry area, 500 m to the south, created Las Rousselon ("red-charred"), a small spatter cone about 10 m high. A larger Strombolian eruption formed the cinder cone of Puy de l'Enfer ("hell"), 200 m farther south, while other eruptions occurred from the southernmost vent, where the main local stream, the River Veyre, crossed the fissure. Its infiltrating waters generated phreatomagmatic explosions that blew out the Narse d'Espinasse, a maar about 700 m across. The vents of Puy de l'Enfer and Narse d'Espinasse were active at the same time, for their fragment layers are interbedded, but the Espinasse explosions were strong enough to destroy the southern half of Puy de l'Enfer – thus neatly illustrating the relative power of the two eruptive styles. Later, when the River Veyre had been diverted for a time by the eruptions, a Strombolian cinder cone grew up within the maar, but the magma stopped rising before it could be built higher than the rim of the maar. Thereafter, the River Veyre resumed its old course and its sediments helped fill the maar, formed the marsh of the Narse d'Espinasse, and buried the cinder cone so effectively that it has only recently been discovered.

Faial, in the Azores, offers several examples of variable environmental influences on basaltic eruptions along fissures extending into the Atlantic Ocean (Fig. 6.17).

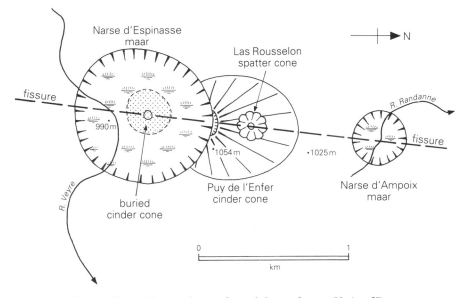

Figure 12.8 Maars and cones formed along a fissure, Chain of Puys.

The western peninsula of Faial is dominated by 16 basaltic cinder cones, called Cabeços ("heads"), commonly over 150 m high, which probably bury older Surtseyan cones (Fig. 6.20). Two eruptions of very similar basalts have been recorded there since the island was settled 500 years ago. The first built up the cinder cones of Picarito and Cabeço do Fogo (the "Head of Fire") and their attendant misterios during a Strombolian outburst lasting from 24 April 1672 until 28 February 1673. The second eruption, off the western prong of the peninsula, created Capelinhos in 1957–8, whose initial Surtseyan characteristics were gradually replaced by Strombolian activity when water could no longer reach the vent.

Eastern Faial illustrates another aspect of environmental influence on basaltic eruptions, where water levels have changed with time along a fissure. On the island, the fissure gave rise to two cinder cones: Monte Carneiro and Monte das Mocas. Where the fissure enters the sea two other volcanoes, Monte Queimado and Monte Guia, erupted at different times. Monte Queimado, as its name "burnt" indicates, is composed of black and red cinders expelled during Strombolian eruptions. But, as its base, and therefore its vent, now lie below sea level, the cone must have erupted on land during a period of lower sea level before being flooded when the waters rose. Its neighbour, Monte Guia ("the signal") is a typical Surtseyan cone, 1 km across, which rises from below sea level to a height of 145 m. Its crater is open to the Atlantic Ocean in the south, and this sector, facing the dominant waves, probably never closed during the eruption. Monte Guia is composed of thin layers of fine yellowish welded tuffs arranged around a broad double crater, formed by two vents operating simultaneously. Monte Guia is less eroded than Monte Queimado and thus erupted later when the Atlantic Ocean was at, or near, its present level, perhaps not long before Faial was settled. Indeed, although Monte Guia is in a rather more protected site, it is very much better preserved than Capelinhos, which has borne the brunt of Atlantic storms for less than 40 years.

13

A volcanic island: Graciosa, Azores

Graciosa is one of the smallest islands in the Azores, but it provides a synopsis of the varied landforms of the archipelago erupted on the flanks of the Mid-Atlantic Ridge. Although many volcanic features on the island are remarkably fresh-looking, no eruptions have, in fact, been recorded since the island was first settled about 1480.

Graciosa forms an ellipse, 12 km long and 8 km broad (Fig. 13.1). Like most of the islands in the Azores, the scenery of Graciosa is dominated by a strato-volcano, which, however, is the only one occupying the southeastern corner of the island (Fig. 13.2). As is also usual in the Azores, Caldeira is the Portuguese name given to the whole volcano – expressing the dominance of the central hollow in the eyes of the early settlers. A wide valley stretching from Praia to Luz separates Caldeira from

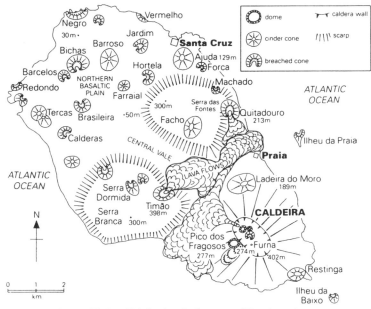

Figure 13.1 Chief volcanic features of Graciosa, Azores.

212

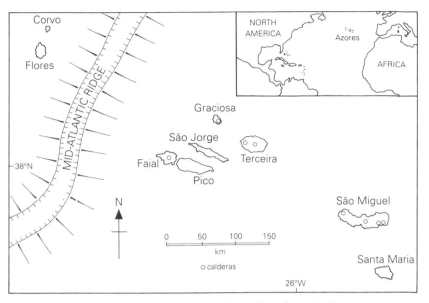

Figure 13.2 The Azores and the Mid-Atlantic Ridge.

the plateaux forming the central, and oldest, parts of the island: the Serra Branca and Serra Dormida in the southwest and the Serra das Fontes in the northeast. These uplands, made up of basaltic lava-flows, face each other across a central vale that widens out into the broad plain, scattered with fresh cinder cones, that covers all the northwestern part of Graciosa. A few cinder cones also crown the Serras and others rise on the flanks of Caldeira itself. The most recent eruption formed Pico Timão cinder cone on the edge of the basaltic plateau of the Serra Dormida. Most of the lava-flows of Graciosa are basalts, with some smaller outcrops of trachytes. Each cinder cone has ejected an extensive skirt of basaltic fragments around its base, whereas the slopes of Caldeira are blanketed with pale trachytic and basaltic fragments stretching as far as the Atlantic coast.

Although the original eruptions that created Graciosa probably began at a depth of 1500m or more on the ocean floor, the most ancient emerged volcanic rocks on the island are basalts under the Serra das Fontes which are about 620000 years old. Trachytes at the base of the Serra Dormida are 350000 years old; those near the summit of the Serra Branca are 270000 years old. All these dated rocks belong to piles of almost horizontal lava-flows, which are beautifully exposed in the red cliffs around the Ponta Branca on the southwestern coast. These basaltic layers reach a height of 375m in the Serra das Fontes and over 300m in the Serra Dormida, where they are covered by lines of much younger cinder cones. The plateau lavas are not only distinguished by their age and nature, they are also clearly separated from the rest of Graciosa by continuous escarpments, 3km long and more than 100m high, facing each other across the central valley and southwards onto Caldeira. These older lavas are probably relics of broader plateaux that were later broken into separate blocks by movements of the Earth's crust.

213

The strato-volcano, Caldeira, covers all the southeastern third of the island, and is its most striking natural feature. Rising only to 402m and stretching only 5km in diameter, the volcano, with its sharp-edged summit caldera, is small enough to be encompassed in a single view. Seen from the Miradouro do Alto Quitadouro on the edge of the Serra das Fontes block to the north, it provides a text-book example of a strato-volcano whose summit has foundered in a great eruption (Fig. 10.2).

The outer slopes of Caldeira are gently concave, increasing from 10° at their base to some 20° near the sharp rim. Its slopes are ribbed by many closely spaced shallow valleys etched only 2m or 3m deep, which rarely form even small bays where they reach the sea. The fringes of the volcano have been eroded to form rugged cliffs, but they are usually less than 20m high, except around Restinga in the far southeast. Fluvial and marine erosion have therefore been only slight, which strongly suggests that Caldeira is a young strato-volcano and that the last events in its life are most probably to be measured in thousands, rather than hundreds of thousands, of years. The strato-volcano was built up by quickly repeated eruptions of basaltic lava-flows and layers of fragments, especially lapilli, ash and tuffs, ranging in thickness from 20cm to 8m. Two thin fossil soils within these layers show that the eruptions were separated by at least two periods of repose, but neither has yet been dated accurately. The mountain may, however, have reached 600m or 750m above sea level before it was decapitated during its climactic eruption.

The caldera is elliptical, 1600m from northwest to southeast, but it is only 875m broad. Its sharp and often ragged rim is also asymmetrical. It rises to 402m in the southeast, and maintains a height of around 350m on its northeastern and south-western crests, but it is notched by a gap that reduces the rim to a height of only 250m in the northwest. The inner walls of the caldera are composed almost entirely of lava-flows, showing that much of the history of the strato-volcano was devoted to relatively calm effusions. The transition to violently explosive eruptions was therefore sudden. The caldera is certainly less than 200000 years old, and its sharp rim, very steep walls and scarcity of weathered material on its surface all suggest that it is unlikely to be much more than 20000 years old. The original amphitheatre may have been more than 400m deep, but it has already witnessed a renewal of activity (Fig. 13.3). In the northwest of the hollow, two basaltic domes rise 100m and already support eucalyptus and laurel. One flank of the western dome has collapsed to reveal its typical onion-skin layering. In the centre of the caldera, three small cratered cinder cones between 50m and 100m high form grass-covered mounds. The most curious feature in the caldera, the Furna do Enxofre ("the sulphurous cave") lies in its lowest part. It is a dark, vertically walled vent, 47m deep, that exhales two vigorous fumaroles and contains an underground lake, whose floor, in fact, lies below sea level. The formation of the Furna no doubt marked the latest eruption within the caldera. However, as the cones and domes were being erupted, a large basaltic lava-flow also escaped through the low notch on the northwestern caldera rim and divided into two branches wrapped around the lower flanks of the strato-volcano. The smaller branch reached the sea near Praia, while the larger mass formed two lobes stretching out into Folga Bay. Because the rim was so low in this

214

Figure 13.3 The floor of the caldera of Graciosa, with wooded domes on the left fore-ground, a cinder cone on the right, and the eastern caldera walls behind.

arca, this is the only case in the Azores where an eruption inside a caldera managed to expel a lava-flow beyond its walls. Until this flow erupted in fact, it is very likely that Caldeira was a separate island.

Meanwhile, other eruptions had taken place on the flanks of the strato-volcano. The Pico da Ladeira do Moro formed an impressive double cinder cone some 150 m high and almost 1 km wide that now dominates the little town of Praia. Close to the southwestern rim of the caldera, Siamese-twin domes form Pico dos Fragosos ("very rocky"), rising to 277 m and 274 m (Fig. 13.4). They have bristling rounded summits composed of rather viscous, almost trachytic, basalt. The eastern dome, 274 m in height, collapsed under gravity as it arose and distended into a viscous lava-flow that has congealed in a 40 m-thick rugged, steep-sided tongue stretching 1 km down slope. The western dome remained intact perhaps because its lavas were even cooler and more viscous. These domes were extruded after the caldera formed because they cover the uppermost layers of trachytic fragments on the surface of the strato-volcano. Other vents, probably also satellites of the strato-volcano, formed the cinder cones composing the islets of Ilheu de Praia and the Ilheu de Baixo. As these are both made up of coarse cinders, they grew up through Strombolian activity on land rather than Surtseyan eruptions at sea. In fact, like Monte Queimado in Faial, they probably erupted when sea level was lower after the last ice age. Both islets, however, could be several thousands of years old, because they have suffered so much marine erosion that far less than half of each original cone now remains.

The northwestern third of Graciosa is a low-lying plain, covering some 25 km^2

215

Figure 13.4   Pico dos Fragosos, Graciosa.

and sloping gently northwards from 55 m to 20 m above sea level. It is composed of a large number of coalescent basaltic lava-flows, often weathered enough to support lava-walled vineyards, meadows and even arable land, which reach the coast in an intricate pattern of low, but very rugged little cliffs that are typical of wave-eroded aa lava-flows. Dotted along generally northwest to southeast-trending fissures, are 20 or so cinder cones, ranging from 50 m to 100 m in height, the most notable of which, Ajuda, dominates the southern skyline of the island's capital, Santa Cruz (Fig. 13.5). All these cones were breached, often almost to their bases, by the basalts that issued from their vents. But the breaches show no relation to steep slopes or preferred fissure or wind directions, and seem to have formed when lava-flows carried away the fragments landing upon them, so that perfect cones never developed.

Several cones, including Redondo ("round"), Negro ("black") and Vermelho ("red") on the northwest coast, have already lost more than two-thirds of their volumes to marine erosion. They are chiefly composed of coarse cinders, but they also contain a few thin layers of yellow tuffs, indicating that they were built partly by Surtseyan eruptions in shallow water. All these coastal cones, and the basalt flows around them, have scarcely been eroded more than Capelinhos in Faial. Inland, too, the basal outlines of similar cones are still sharply defined and have been little disturbed by river erosion or weathering. It seems likely, therefore, that all these cinder cones and their lava-flows can only be a few thousand years old at most, and could have erupted only just before the island was settled about 1480.

The fissures which gave rise to the eruptions on the northwestern plain of Graciosa were also activated at about the same time on the ancient lava plateaux, when about 15 cinder cones were formed and most of them were also breached. Here, though, relief has played an important rôle in their growth, for several cones erupted from vents on top of the escarpments around the edges of the plateaux. Quitadouro

216

cinder cone, erupted on the edge of the Serra das Fontes, could not build up the sector of its cone that sloped steeply southwards towards Praia. Its lava-flow and slumped cinders now make a chaotic tongue issuing from the much-quarried reddish-brown armchair of Quitadouro. Its counterpart, Pico Timão, is only 100 m high, but its base on the edge of the Serra Dormida already approaches 300 m above sea level, so that its summit occupies a commanding position, at 398 m, overlooking most of central Graciosa. Timão forms a crescent of bare reddish cinders that is breached open facing the escarpment where it erupted. As at Quitadouro, Timão could not maintain this sector as lavas welled up, carried away erupted fragments and broadened the breach by pushing aside the arms of the crescent. The basalts flowed into the valleys fringing Caldeira, and entered the sea, 4 km away, in a lobe 1 km wide north of Praia. The flow covers the snout of the Quitadouro basalts as well as those which issued from the caldera during its latest spasms. This was the most recent eruption on Graciosa and the cone still looks fresh, with bare cinders and a clearly marked crater rim and basal outline. The flow has maintained its scarcely weathered, rugged aa surface and edges, which plantations of eucalyptus, cryptomeria and laurel cannot disguise. Its snout, also, juts out like new into the sea and has hardly been eroded by the waves. The appearance of Timão cone and its flow strongly indicates that they cannot have erupted long before the first colonization of Graciosa about 1480. Indeed, most of the surface features of the island are so geologically and morphologically youthful that it is surprising that no eruptions have been recorded during historic time. The past five centuries of calm seem to be a quirk, perhaps not truly representing the nature of activity on Graciosa during the past few thousand

Figure 13.5 Ajuda cinder cone, Santa Cruz, Graciosa.

217

years. It is likely that not very many centuries will elapse before the island emulates most of its neighbours and sees another cinder cone erupt.

14

Volcanoes and erosion

Volcanoes do not remain intact for very long after they stop erupting because atmospheric agents of erosion soon attack them and alter their appearance. These agents, chiefly rivers, weathering, glaciers, landslides and the sea, are all very selective in their action. They attack weaker rocks faster than stronger rocks, so that the weaker are carved into valleys and the stronger survive as hills and ridges. Volcanic rocks are particularly prone to selective erosion because they are so varied. In general, the compact rocks such as lavas occurring in flows, domes, or vent plugs are the most resistant to atmospheric agents, whereas loose fragments that are not bound together with a firm cement, such as ash, tuffs, lapilli, cinders and pumice, are always weak. One exposed layer of loose rock can weaken a whole mass so that it is quickly degraded, but, on the other hand, a lava-flow can also form a protective cap-rock which may delay erosion for millennia. Thus, a volcano composed of many lava-flows is much more resistant to atmospheric degradation than one that has been built up of unconsolidated fragments. Lava is the armour-plating of volcanoes.

Marine erosion

The waves attack and undermine the superstructures of volcanoes along the nearly horizontal plane of sea level. The eastern flank of the first cone of Vulcanello, which was born in 183 BC, has already been so deeply eroded by the sea that the details of its internal layers have been laid bare (Fig. 14.1). At Strombolicchio, which arose from a flank vent of Stromboli, the sea has completely removed all the weak cinder layers and only the old lavas plugging the vent remain like a ruined castle.

Surtseyan tuff rings and tuff cones are particularly vulnerable to marine erosion, because they are composed of layers of weak fragments (Fig. 14.2). Most of the islands created solely by Surtseyan activity have remarkably brief lives, like the two islet tuff cones formed on the flanks of Surtsey. Syrtlingur, born on 5 June 1965, reached a height of 65 m before being destroyed above sea level by a week of storms beginning on 17 October 1965. Jolnir islet fared little better. Emerging on 28 December 1965, it reached a height of 60 m by 10 August 1966, but marine erosion gained the upper hand and removed it above sea level by 20 September 1966. Neither islet was protected by lava-flows.

Figure 14.1   The eastern flank of Vulcanello, Vulcano.

Faced with marine attack, massive lavas are also vulnerable, primarily because their well developed vertical columns are ideal for cliff development. The lavas are undermined at their base and the whole column then collapses into the sea, leaving

Figure 14.2   Eroded tuff cones at Cap d'Agde, France.

220

behind a colonnade, which is attacked in its turn. These features have been made famous by Fingal's Cave at Staffa in the Scottish Hebrides and by the Giant's Causeway in Northern Ireland. Lavas about 250000 years old from Etna on the shores of Sicily near Acireale have been eroded into picturesque headlands and islets known as the Riviera degli Ciclopi, so named because Homeric legend has it that they were thrown at Odysseus by Polyphemus, the tormented Cyclops.

The coastal fringes of Sete Cidades volcano on São Miguel illustrate some of the effects of the constant battle between volcanic growth and marine counter-attack in the past 5000 years (Fig. 11.3). The lavas of the periphery of the volcano itself have been eroded by the waves to such an extent that a cliff about 500 m inland and 100 m high now borders the coast. On its flanks, Pico das Camarinhas retaliated successfully about 840 years ago, when it erupted the flow that made the lava delta of the Ponta da Ferraria (Fig. 14.3). On the other hand, off this promontory between 1 February and 14 July 1811, a Surtseyan eruption built up the islet of Sabrina, which survived for less than a month as soon as the eruption stopped. Nowhere on São Miguel, perhaps, are the interactions between the sea and volcanic activity better displayed than around the village of Mosteiros on the northwestern coast. It lies on a flat-topped lava delta, formed when basalts from a satellite vent cascaded down the cliff bordering Sete Cidades volcano. The original surface roughness of the lavas was planed off by the waves when sea level stood about 15 m above that at present. Their coastal edges, however, are now extremely rugged where the waves have picked out the weaker, cindery layers between the basalts. Offshore stand the four islets comprising the Ilheus dos Mosteiros, which are all that remains of a Surtseyan cone. Off

Figure 14.3 Pico das Camarinhas cinder cone and its lava-flow extending into the sea, São Miguel.

221

the northern end of the village, an older cinder cone can only be detected in the intertidal zone by the concentric outcrops of black spatter. It is the oldest feature exposed on the coast and had already been eroded down to sea level when the lava delta on which Mosteiros stands covered its southern half. Neither the eruptions nor the sea are clearly winning their battle on São Miguel at present.

Glacial erosion

Ice adds to the glamour of many volcanoes by crowning their summits, but also contributes to their destruction, for a high volcano, with its low temperatures, abundant snowfall and steep slopes, is the ideal place for glaciers to develop. Glacial erosion soon carves deep troughs into the weak ash, cinders or pumice in many strato-volcanoes. Even massive lavas are not immune: they merely withstand erosion longer. Glaciers have deeply carved Cantal volcano from a focus at the dome of Puy Mary. The ice quarried and scalloped the dome on all sides to such an extent that it lost half its original volume, and it is now a fine example of a glacial horn, or pyramidal peak (Fig. 14.4).

Mont-Dore volcano offers an even more striking illustration of how the intensity of glacial erosion varies within a small area. Its eastern flanks were gouged out by glaciers to form the 1000m deep trough of the Chaudefour valley. On the northern slopes of Mont-Dore, another glacier first carved only the shallow depression now filled by the Lac de Guéry, but then plunged over a lava step some 200m high, squeezed between the phonolitic pitons of Roche Tuilière and Roche Sanadoire, and eroded a wide U-shaped trough extending 15km farther north, mainly through layers of pumice capped by thin lava-flows (Fig. 8.2). Just to the northeast, on the

Figure 14.4   Puy Mary, Cantal volcano, France, a glacially carved dome.

other hand, a weaker, shorter, glacier merely smoothed the basaltic lava plateau near Aurières. However, the intensity of glacial erosion also depends on the pre-glacial landscape. The Chaudefour valley, for instance, had already been deeply carved by the River Couze Chambon, whereas the Aurières plateau had scarcely been touched by rivers when glaciers came onto the scene.

Landslides

Many landslides are caused by volcanic activity, but these, like debris avalanches, are best considered as the direct results of eruption. The landslides that take place after the end of activity occur on many different scales. Rock-falls are common within steep-sided craters; gravity collapse commonly occurs on many new cinder cones and tuff cones; and rockslides are frequent on scree slopes. The Rina Grande, on the flanks of Stromboli, is a large, unstable scree (Fig. 5.8). It may well be, also, that the Sciarra del Fuoco on the opposite side of Stromboli began life as such a rockslide before becoming the routeway for lava-flows sliding seawards.

Larger landslides can occur where lava-flows lie above water-holding rocks. The Trotternish peninsula in Skye in Scotland provides the two most imposing examples in Britain in the Quirang and the Storr landslides. The basaltic flows, which are permeable because of their pronounced vertical jointing, lie above impermeable water-holding clays. Both tilt eastwards towards the Sound of Raasay, where they were eroded and destabilized by the waves and by glacial gouging during the Ice Age. As a result, the basal support of the lavas was eroded away. The whole mass thus became unstable and the lavas slipped under gravity down the lubricating clay layer like a sledge moving over the snow. Enormous masses of lava were detached from the Trotternish peninsula, leaving behind great arc-like scars on its crest. As they slipped, the lava-flows broke into huge, chunky blocks forming the chaotic maze of the Quirang. One lava column, called the Old Man of Storr, was detached from its parent flow and now stands guard at the southern end of this disturbed landscape.

Similar landslips have occurred along the Columbia River gorge in the United States, where layers of basalt and lubricating layers of deeply weathered ancient soils are arranged, as one graphic phrase has it, like a southward-sloping stack of buttered pancakes. The Columbia River has undermined their base, and many landslides have developed, notably where the lavas tilt river-wards near North Bonneville.

Landslides on an enormous scale have occurred on both Etna and Piton de la Fournaise in Réunion, and for apparently similar reasons. The massive bulk of Etna is supported on its western flanks by the sedimentary rocks forming most of the rest of Sicily, but its eastern flanks drop steeply to the floor of the Ionian Sea. As a result, the eastern slopes of the volcano are sliding intermittently seawards. In Réunion, Piton de la Fournaise is supported on its western side by Piton des Neiges, but its eastern flanks fall away to the Indian Ocean (Fig. 5.7). Here, too, the eastern slopes are sliding seawards, forming the Grand Brûlé, which on land alone is 13 km long and over 7 km wide. It originated after the Enclos Fouqué caldera collapsed about

4745 years ago, and has been sliding down piecemeal fashion ever since, so that it now extends far below sea level.

Fluvial erosion

Streams are by far the most important agents of erosion on nearly all landscapes. They are also the most selective in picking out minor variations in rock resistance. Nevertheless, if they were left to work unhampered, they would not take many millions of years to reduce a large strato-volcano to a mere stump, from which only the most resistant lavas would protrude. The destruction of a small cinder cone, of course, requires much less erosional effort.

Active cinder cones and tuff cones usually have smooth straight slopes, but the rains soon develop gullies which radiate from the crater rim and form typical parasol ribbing that is well illustrated on Caldeira Seca in São Miguel. Eventually the weak ash or cinders are transported away, so that only lavas in flows or vents then remain. Vent plugs, for example, make Le Puy one of the most striking towns in all France (Fig. 8.6). These pinnacles are all that survive from hydrovolcanic features formed about two million years ago. Rocher Corneille is an elliptical pinnacle, 130 m high, with the Cathedral at its foot; Rocher St Michel is an elongated pyramidal needle, 82 m high, that is crowned by a medieval chapel (Fig. 14.5). The Devil's Tower in Wyoming and Ship Rock in New Mexico are remnants of similar lava-plugged vents in the United States (Fig. 14.6). Because they extend vertically down through the crust, such plugs are likely to remain as landscape features for millions of years. Those in central Scotland, forming the craggy bastions of Stirling and Edinburgh Castles and the "Laws" at Dundee, Haddon and Largo, for instance, still stand out as prominent hills, although they are 250 million years old.

Strato-volcanoes are composed of a greater variety of materials than cones, and streams thus act upon them in more complicated ways. Most strato-volcanoes are gullied by streams radiating from their summits. These are often known by the Spanish term "barranco" and they can be seen, for example, on the slopes of Teide, Agua and San Pedro (Fig. 10.4). As streams carve their valleys, they become increasingly selective in removing weaker rocks; the barrancos develop tributaries and also cut gorges deeply into the flanks of the volcanoes, especially where they are unprotected by lavas. The summits of many strato-volcanoes are vulnerable because they are steep and often composed of weak ash, whereas their lower slopes, sometimes known as "planèzes", often resist better because they are gentler and often armour-plated with widespread lava-flows of varying ages. The central cores of many ancient volcanoes have thus been gutted by erosion. Such is probably the origin of the great hollow at Curral das Freiras in Madeira, and of the Taburiente "caldera" on La Palma in the Canary Islands. One of the most spectacular cases occurs on Piton des Neiges in Réunion, where the volcano has been excavated by three separate drainage systems, which have cut out three vast pear-shaped hollows: the Cirque de Mafate, the Cirque de Cilaos and the Cirque de Salazie (Fig. 5.7).

Figure 14.5   Rocher St Michel, Le Puy, France.

Cantal volcano has been so degraded by streams (aided at times by ice) that its summit caldera has vanished as a landscape feature. Its crest is now a meandering ridge of resistant rocks, linking Plomb du Cantal with Puy Mary, that has been carved out of the old caldera floor.

Erosion in the Cascade Range

The Cascade Range of the United States and Canada is a series of recent strato-volcanoes, all erupted within the past 700 000 years, which are now in varying stages of activity and erosional decay (Fig. 6.5). The active strato-volcanoes have maintained the smoothest conical outlines, whereas their extinct and deeply eroded companions are so jagged that they no longer resemble cones at all. Before 1980 Mount St Helens was certainly the smoothest and most graceful cone in the whole Cascades, but now, perhaps, that distinction would fall to Shastina on Shasta. On the

225

Figure 14.6   Ship Rock in the distance and a lava-filled dyke in the foreground, set in relief by erosion, New Mexico.

other hand, Thielsen, Mount Washington and the aptly named Three Fingered Jack are old strato-volcanoes that have been reduced by streams and ice to jagged, crumbling lava stumps. Thus, a shattered feeder dyke now marks the crest of Three Fingered Jack, and andesitic lava plugs the vents now forming Thielsen and Mount Washington. Mount Hood, Shasta and South Sister fall between these two extremes and have maintained their broadly conical shape while also showing some atmospheric decay (Figs 10.1, 10.7, 10.8). It is tempting, as with human beings, to judge their age by their appearance. Sometimes it works. Thus, amongst the Three Sisters volcanoes west of Bend in Oregon, North Sister looks the oldest, followed by Middle Sister, and South Sister looks the youngest (Fig. 14.7). Appearances are not deceptive: the relative ages can be confirmed. North Sister has lost its crater and perhaps a quarter of its volume to the depredations of glacial erosion during two ice ages, leaving behind two deep glacial cirques on its northern and eastern sides. Middle Sister apparently suffered only the last episode of glacial erosion, which excavated part of its eastern sector. Both the summit cinder cone of South Sister and its rather older flanks remain virtually intact because they had not even been formed when the last ice age ended. Moreover, although its older siblings are no doubt already extinct, South Sister is probably still active, and saw the eruption of rhyolitic domes on its flanks between 2300 and 1900 years ago.

Other Cascade strato-volcanoes often show greater recent erosional destruction than volcanic construction. They were particularly vulnerable to the effects of the ice ages because they accumulated great depths of snow that changed into glaciers which gouged their crests and, when the ice periodically melted, they suffered land-slides, mudflows and stream incision as well. Thus, only those peaks which have erupted frequently since the last glacial period ended about 10000 years ago have preserved their pristine outlines almost intact. Until 1980, Mount St Helens was the chief amongst these. Elsewhere, erosion has taken a considerable and varied toll.

Part of Mount Garibaldi, in British Columbia, was constructed during the ice ages over an ice-filled valley. When the ice melted, the western half of the original cone slid in sharp bursts into the nearby Squamish valley, considerably steepening the slopes of the remnant in the process. The upper northern sides of Mount Hood, Jefferson and McLoughlin in Oregon have been eroded by ice and depleted by land-slides and lahars, so that each has lost its crater and probably between 150m and 300m in height. In central Oregon, Mount Bachelor and South Sister are almost intact cones, but nearby Broken Top, as its name implies, has a jagged, ruddy crest that has been excavated to its core by streams and glaciers (Fig. 14.8). The most spectacular recent losses of all have been incurred by Mount Rainier, which still, however, remains the largest volcano in the Cascade Range, rising to a height of 4394m. It still carries 26 named glaciers, which, of course, were much larger in the ice ages. They have carved deep troughs, such as the Nisqually and White River valleys, into its flanks, and each valley radiating from its summit is separated by nar-row ridges that, perhaps, constitute the sole surviving remnants of its original slopes. But the rocks attacked varied in strength and the glaciers excavated valleys by no means uniformly into them, so that each side of Mount Rainier looks different. Not only was Mount Rainier deeply scoured by glaciers, but it lost its summit cone in a great landslide, possibly triggered by an eruption about 5700 years ago. The cone, mixed with summit ice, quickly changed into the Osceola lahar which swept down the White River valley and redistributed the summit of Mount Rainier over the Native American settlements near Puget Sound 100km away. Mount Rainier became about 500m lower in the process. This lahar was the largest of several which have degraded the volcano since the last ice age ended. Eruptions have thus failed to keep pace with erosion even on Mount Rainier for several thousand years at least.

Figure 14.7 Three Sisters, Oregon: South Sister on the left and North Sister on the right.

Erosion and lava-flows

Lava-flows often have a great influence on erosion. They have a dual rôle because they are eroded by streams and yet often direct their courses, and can sometimes even change the whole balance of selective erosion in a volcanic area. Erosion may also depend on the amount of lava erupted and the nature of the land surface it invaded. A lava-flow can enter a valley and block the main stream, as occurred in Auvergne, when the Cheire d'Aydat impounded the River Veyre and held back the waters of the Lac d'Aydat (Fig. 14.9).

Some lava-flows are so thin that they merely form ribbon-like strips hugging narrow valley bottoms. Their position makes them always vulnerable to stream attack and they are usually quickly dismembered. The lava fragments may, indeed, accelerate erosion because they provide the stream with hard abrasive tools. Such a ribbon-like flow in the River Auzon valley in Auvergne has already been broken up, although it only erupted 60000 years ago (Fig. 14.9).

When vast fluid basalts accumulated to form the Columbia volcanic plateaux, the earlier landscape was completely covered and new drainage had to start afresh. Only powerful streams such as the Columbia River and the Snake River could maintain their courses in the face of such swamping. They have carved deep, steep-walled gorges because the strong vertical jointing of basalt makes it a most suitable medium for stream incision. Their tributaries cut back great waterfalls such as Multnomah Falls that cascade into the Columbia River gorge.

What was probably the fastest and amongst the most extraordinary episodes of erosion was accomplished by glacial meltwater on the Columbia volcanic plateaux at the end of the last ice age. As the ice-sheets melted away in Montana, masses of water became trapped in lakes between the melting ice and the surrounding hills. The greatest of several such lakes formed at Missoula in Montana between about

Figure 14.8 Broken Top and South Sister, Oregon.

228

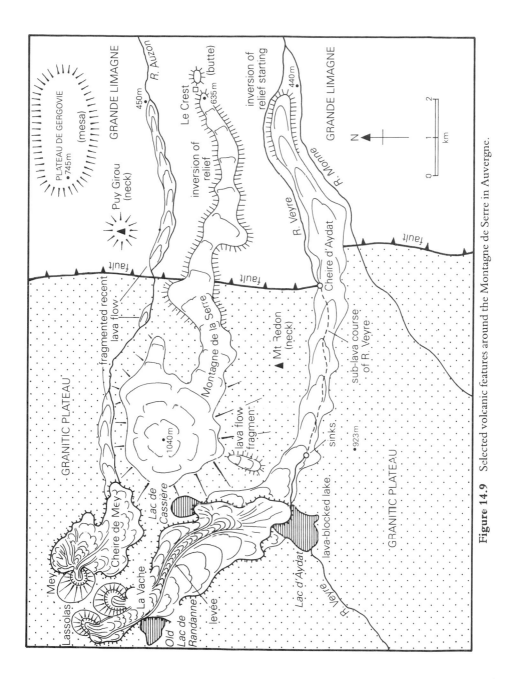

Figure 14.9 Selected volcanic features around the Montagne de Serre in Auvergne.

Labels within the figure:

GRANITIC PLATEAU

fragmented recent lava flow

Mey
Cheire de Mey
Lassolas
La Vache
Old Lac de Randanne
levée
Lac de Cassière
•1040m
Montagne de la Serre
lava flow fragmen—
▲ Mt Redon (neck)
sub-lava course of R. Veyre
sinks
•923m
R. Veyre
lava-blocked lake
Lac d'Aydat
Cheire d'Aydat
fault
GRANITIC PLATEAU

fault

Puy Girou (neck)
▲
PLATEAU DE GERGOVIE
•745m
(mesa)
GRANDE LIMAGNE
R. Auzon
450m
Le Crest
•635m (butte)
inversion of relief
inversion of relief starting
R. Veyre
R. Monne
440m
GRANDE LIMAGNE
fault

N

0   1   2
km

Figure 14.10   Dry Falls, Washington State.

12000 and 13000 years ago. Water accumulated to a depth of 300m in Lake Mis-soula until it suddenly caused the ice dam to fail. In less than three days, a mass of water equivalent to half the present volume of Lake Michigan rushed out at a rate which was 2000 times the discharge of the present Columbia River. The flood swept across the volcanic plateau in southeastern Washington and carved out a be-wildering array of channels in the basalts known as the "scablands" (Fig. 14.10). Its most impressive results are the Grand Coulee canyon and Dry Falls, farther south, which, for a few days, were 100m high and 5km wide – a veritable Niagara, and three times its size! It is important, however, to remember that lava-flows are not usually eroded with such efficacy in such a brief catastrophic interval . . .

Volcanic inversion of relief

The really interesting changes arise, however, when enough lava is erupted to divert streams and dominate the drainage pattern without completely altering it. Lavas fill the valley bottoms but are not thick enough to cover the ridges between the valleys (Fig. 14.11). The stream that used to occupy the valley then divides, so that twin streams run alongside the edges of the flow. They carve new valleys alongside the old valley that is protected and fossilized by the resistant lavas. The old unprotected ridges are gradually eroded away and carved into valleys, with a new ridge between them along the lava-flow. Thus, volcanic inversion of relief occurs because a ridge develops in place of the old valley now protected by lava, and two valleys replace two former ridges.

Volcanic inversion of relief can create some striking landforms. The lavas of Rampart Ridge on the southwestern flanks of Mount Rainier near Longmire, for example, have been set in relief by erosion of weaker cinders alongside them. The Montagne de la Serre in Auvergne is a famous case of volcanic inversion of relief (Fig. 14.9). Its lavas erupted three million years ago, flowed eastwards from the granitic basement of the Chain of Puys and entered a valley leading to the Grande Limagne, where they covered weak sands. The River Allier and its tributaries then excavated the Grande Limagne so that its level where the weak sands were unprotected was lowered by more than 200m. But the lavas formed a strong shielding arm over their old valley and prevented the weak sands lying beneath them from being eroded, so that the Serre, as the French name, "narrow ridge", implies, now forms a lava-capped ridge, 200m high, jutting out eastwards into the Grande Limagne as far as the village of Le Crest. The lower part of the Montagne de la Serre in the Grande Limagne thus demonstrates an inversion of relief of at least 200m.

The Cheire d'Aydat, 2km to the south, shows what the Montagne de la Serre must have originally looked like. This lava-flow erupted from Puy de la Vache and

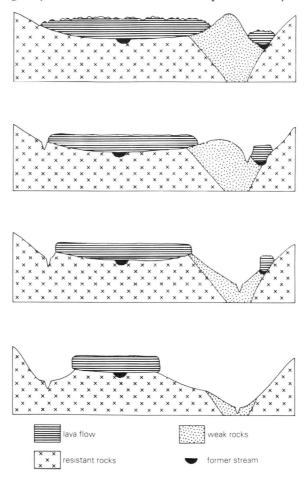

Figure 14.11   The development of volcanic inversion of relief.

231

Puy de Lassolas about 8000 years ago and runs almost parallel to the Serre into the Grande Limagne. The lavas of the Serre must have filled a similar valley three million years ago. Although the Cheire d'Aydat lava-flow is young, its own snout has already started to undergo inversion of relief because both the River Monne and the River Veyre alongside it have cut their beds about 10m below the level of the valley that the lavas first protected only 8000 years ago.

Although they protect weaker rocks, the lavas causing volcanic inversion of relief are no more immune to attack in the long run than any other rocks. The edges of the flows are fretted until the internal lava columns are revealed and are themselves undermined so that the whole face collapses and exposes another vertical cliff to attack. The flow is reduced until smaller lava masses are isolated. For example, the snout of the Montagne de la Serre has already been separated from the main ridge to form a butte at Le Crest (Fig. 14.9).

Lava-covered ridges and plateaux may be fragmented into smaller units which often crown conspicuous flat-topped hills called "mesas", from the Spanish term for table. The Plateau de Gergovie is a famous mesa capped by resistant basalts filling two maars, which was steep-sided enough to be a last redoubt of the Gauls against Julius Caesar (Fig. 14.9). Eventually mesas are themselves reduced to smaller "buttes" and to thin ridges, called "cuchillos" in Tenerife. Occasionally, lava-flows that once flooded into valleys remain for a time, like the Montagne de la Serre, as elongated mesas. Their presence in the landscape then enables old drainage patterns to be reconstructed. For example, such relict features now form basalt-capped mesas at Bergonne in the Grande Limagne, which have fossilized an old course of the River Allier. All these features stand out in the landscape as long as the lava cap remains, but, once the cap-rock is removed, they are soon eroded away – unlike lava-plugged vents which last much longer.

Rather special landforms are sometimes created by the selective erosion of ash-flows which have been welded to varying degrees. Nowhere, perhaps, is the result of ashflow erosion so beautiful as in Cappadocia in Central Turkey (Fig. 14.12). Unwelded yellow pumice gives spectacularly gullied badlands, whereas a welded cap-rock produces broad surfaces sloping smoothly from the parent vents. Intermittent welding, however, in and around the villages of Avçilar, Ürgüp, Cavuş In and Ortahisar, forms a landscape of buff and yellow pinnacles of pumice, varying from 5 to 50m in height and enough to dwarf the village minarets. The pumice has been excavated for dwellings, storage barns and even old Greek Orthodox churches. The larger masses form rugged towers once used for defence, as the name of the village of Ortahisar indicates, and smaller masses give pinnacles capped by narrow cones in the shape of penitents' hoods. It is one of the world's most bizarre landscapes.

Atmospheric agents acting on varied volcanic rocks thus produce landforms ranging from featureless lava plateaux in eastern Oregon to the strange pinnacles of Cappadocia. They also weather volcanic rocks into thick and fertile soils, whose high yields tempt farmers to cultivate land – perhaps too closely – to potentially dangerous volcanoes.

Figure 14.12   Pinnacles of eroded pumice, Avçilar, Cappadocia, Turkey.

PART IV

Prediction

15

Predicting volcanic eruptions and their effects

All those who live on or near an active volcano are making a volcanic prediction. They may not think about it in those terms, but they balance the benefits of staying near a volcano against the chances of losing their lives or jobs. They predict that they, and their property will survive. If, however, they live on materials erupted from an active volcano, they run the risk that a similar eruption will overwhelm them in the future. The longer their homes are spared, the more confident the people will become that they are safe — and the more likely it is, in fact, that they will be proved wrong, because active volcanoes that have been quiet for a long time have a way of breaking their silence by cataclysmic explosions with little warning. But it is hard to predict when an eruption will take place, to assess its likely power, and to calculate the risks facing the local population. In some countries the inhabitants scarcely have a choice. Many towns in Japan and Indonesia, for example, are too close to explosive volcanoes. It is hardly surprising, then, that these countries account for two-thirds of the deaths caused by volcanoes since 1600. Safe choices are not always made anyway. People often knowingly put themselves in danger, which they clearly perceive as small. They fall prey to the gambler's fallacy: that an event is unlikely to repeat itself in the same place. Catania was rebuilt on the lava-flows from Etna that had destroyed most of the city in 1669. St Pierre was rebuilt over its own ruins soon after the eruption of Montagne Pelée in 1902. Although the risks were re-emphasized by the eruptions in 1929–31, St Pierre now has 7000 inhabitants.

Prediction in perspective

Predicting volcanic eruptions is not a game, designed to back a winner in some kind of volcanic horse-race. It is applied volcanology, to protect and improve the quality of life of the local inhabitants, and to limit the damage and deaths that may be caused. Some kinds of prediction are easy. Amongst the Italian volcanoes, Stromboli will erupt within an hour from now, because it has performed several times an hour for most of the past 2000 years. Etna, which has been bustling with activity since

1971, might continue to do so in the near future, although the long, voluminous emission of 1991–3 might, like its predecessor in 1669, start a period of tranquillity. Vulcano, whose outbursts occur about once every 100 years on average, and has been calm since 1890, might well erupt before the present century ends. Vesuvius might not revive soon, because it seems to be starting one of its dormant periods. However, such generalizations (where "might" occurs four times) would not surprise an expert because they are based on well known trends. Moreover, the forecast about Vulcano, for example, would obviously not justify its evacuation tomorrow – and who would take the responsibility for removing summer tourists on such evidence?

Generalizations based on broad trends are thus of limited use, because little preventive action could be taken about them. If, also, a volcano behaves in a consistent way, like many active cinder cones, for example, then it is often not very dangerous and its behaviour can be readily assessed. After all, the expected scarcely needs a forecast. In such cases, the people keep their distance – and usually know what distance to keep. It is, however, the most violent and dangerous volcanoes that display the most irregular behaviour, and have erupted most lethally, and often without apparent warning in the past. Forecasting the unexpected is thus the real problem. This requires more sophisticated analysis, which has been developed with increasing success, so that volcanologists are becoming more ambitious about predicting eruptions. The study is still, however, in its infancy and problems remain. This is a pity because not only could lives be saved, but Earth scientists might also earn high fees as consultants.

Prediction has to be set in a broad perspective. The terms "active", "repose", "dormant" and "extinct", that are so often applied to volcanoes, still cannot be clearly defined. An active volcano need not necessarily be in eruption, just as someone with a cough does not cough all the time. Eruptions are separated by periods of repose, which eventually last for so long that the volcano is considered dormant and eventually extinct. Stromboli and Etna are active; Vulcano and Vesuvius are dormant; Fuji could be dormant or extinct. Different volcanoes, however, experience dormant periods that can last for decades or centuries. Vesuvius, for instance, presents challenging problems, because it has been the most violent volcano in Europe in historic time, because it lies in a densely populated area, and because its present state of activity is uncertain. The interval since Vesuvius was last active in 1944 is now longer than any since the eruption of 1631 ended several centuries of repose, and it seems certain that its next revival will be greater than any other European volcanic outburst since AD 79. It is also hard to know whether a volcano is truly extinct, and some eruptions have caused great surprise even in the present century.

The other aspect of broader perspective concerns the perception of volcanoes as purveyors of destruction. Most volcanoes do not threaten very often. Only about 1 per cent of the world's eruptions during the past 100 years have caused any fatalities, and there are about 30–50 eruptions on land throughout the world in a year, lasting an average of 6–10 weeks. Luckily, the large Plinian explosions are witnessed only about 10 times a century. About 70000 people have died as a result of volcanic eruptions since 1900, and more than half those deaths occurred in a few lethal moments on two days: at St Pierre on 8 May 1902 and at Armero on 13 November 1985. Almost 200 000 people died from volcanic activity in the 19th century, but two-thirds of those deaths were caused in Indonesia by Tambora in 1815 and Krakatau in 1883. The "average risk" – if such a term has any meaning – to any individual from an eruption is thus very small. Hurricanes and earthquakes are far more lethal, and the threat is altogether greater from delights such as alcohol, cigarettes or motor cars. But an eruption is strongly localized in both space and time, and a calamity impresses more than the banality of deaths from more readily accepted causes.

Volcanoes are thus perceived as being much more dangerous than they really are. These exaggerated perceptions, however, frequently outweigh the reality, as Pliny the Younger noted at Misenum. The panic in Guadeloupe when the Soufrière erupted in July 1976 was not based on its own previous behaviour, but on the violent reputation of Montagne Pelée. All volcanoes are commonly assumed to behave in the same way, and even cinder cones are sometimes believed to be as dangerous as strato-volcanoes. But danger varies even amongst strato-volcanoes. Although Vesuvius has killed thousands, Etna and Stromboli, for example, cannot have caused 500 fatalities for all their continued activity in historic time. Thus, these misconceptions are almost as important as the truth when the correct preventive measures have to be taken.

The final and most important of all prediction perspectives is that no volcanic eruption can be stopped. Its effects – at best – can only be alleviated.

Long-term predictions

There are two kinds of volcanic predictions: long-term forecasts on the basis of historical and geological records; and short-term forecasts based on surveillance.

Study of the present behaviour of any volcano is of paramount importance because the more dangerous specimens can then be selected for closer observation. Knowledge of a volcano's behaviour is clearly increased if it has been studied for a long time. Written records of the Mediterranean and Japanese volcanoes stretch back more than 2000 years, although the earlier accounts are often entangled with myths and legends. On the other hand, Native American stories represent the only references to eruptions in the Cascade Range before the close of the 18th century. In any case, the volcanic record of the past few centuries in a life of perhaps more than half a million years may not be at all typical. Historical records, then, are often too brief, ambiguous and incomplete to provide a clear pattern.

The geological record thus remains the most valuable resource for establishing the true, long-term behavioural characteristics of volcanoes. Geologically determined prediction is based on the assumption that the past is the key to the future. Eruptions are sometimes revealed in the geological record that would be quite unexpected from study of the historic past. Thus, several calderas have formed on Etna during the past 50 000 years although it is usually much less violent. Montagne

Pelée has had only Pelean phases in 450 years of historic time, but its geological record reveals episodes of Plinian activity, which, if resumed, would endanger over 30000 people in an area five times larger than that devastated in 1902. A careful assessment of volcanic risks was undertaken in the Cascade Range in 1975 that did much to make up for the lack of historic data there. As a result, Mount Baker, Mount Rainier, Mount St Helens, Mount Hood, Mount Shasta and Lassen Peak were identified as potentially dangerous. Mount St Helens was singled out as the most explosive volcano in the Range, and, in a triumph of precision, it was forecast that an eruption was likely to occur before the end of the century, although it had hardly been more active than most of its companions in historic time. On the other hand, the forecast in 1983, based on apparently firmly established trends, that Hekla, in Iceland, was unlikely to erupt before the next century, was confounded by its revival in 1991. But a less categoric forecast, also made in 1983, that an eruption in São Miguel was "considerably overdue", still remains unfulfilled.

The testimony derived from geological studies also indicates the kind of activity that usually prevails, and therefore what materials may be expelled. Obviously the behaviour of a cinder cone eruption is much easier to forecast than that of a strato-volcano, and the effects of lava-flows may be assessed far more accurately than those of Plinian ashfalls. The geological record of the Soufrière in Guadeloupe, for example, shows that relatively harmless phreatic ejections are ten times more likely to occur than dangerous nuées ardentes. It is inconvenient, though, that many of the most violent eruptions leave little trace in the geological record. Blasts and ground surges, for instance, only leave thin and inconspicuous deposits and their importance as danger indicators is only just being realized. Moreover, many violent explosions produce such fine and loose fragments that they are quickly removed by erosion. Pumice, a notable indicator of dangerous activity, will even float away. Some dangers may not even be envisaged because they have either never been observed or correctly interpreted. Such was the case with the nuées ardentes of Montagne Pelée in 1902 and the blast and debris avalanches at Mount St Helens in 1980. But, as soon as they are recognized, risks can then be reassessed on other similar volcanoes. Geological studies can highlight the volcanoes which are potentially the most dangerous and should be selected for careful monitoring. In this way, limited resources can be channelled in the right direction, especially in poorer countries often prone to more pressing natural hazards. Perhaps the main limitation of prediction based on studies of geology is that the exact nature – and especially the date – of the eruption cannot be furnished. It is a snag that can only be overcome by continual surveillance.

Danger assessment

Predicting an eruption is not the same as predicting a danger. Danger assessment pin-points the nature and frequency of activity in relation to the local relief and to the vulnerable population near the volcano. A small cinder cone eruption is rarely hazardous more than 1 km away, and even a large eruption may not necessarily cause casualties. The extremely powerful outbursts of Bezymianny in 1956 and Katmaï in 1912 caused widespread devastation, but presented no danger to human life because they only affected unpopulated areas. Similarly, in historic time, 95 per cent of all the eruptions on Piton de la Fournaise in Réunion have occurred in the uninhabited Enclos Fouqué caldera. But an eruption near a large town can be much more dangerous. Towns on the Bay of Naples, such as Torre del Greco, for example, have been destroyed time and again by lava-flows from Vesuvius.

Different types of activity present different potential threats. Fast-moving, sudden, ground-hugging or wide-spreading products of eruptions, such as ashfalls, jökulhlaups, lahars, toxic gases or nuées ardentes, are obviously the most dangerous and destructive. A nuée ardente killed 28000 people in St Pierre in two minutes, the inhabitants of Nyos village were gassed without warning, and Pompeii was buried in a day, but the lava-flow that erupted in 1669 on the flanks of Etna took five weeks to travel 12km to the walls of Catania. The low volume and discharge of most Icelandic fissure eruptions mean that they usually present but a small threat, especially since they often occur in sparsely populated areas. It is possible to walk away from a solidifying lava-flow, but a nuée ardente can travel at 500km an hour. In the present century alone, it was nuées ardentes, allied sometimes to great ashfalls, that were responsible for the deaths of 28000 people at Montagne Pelée, 1565 at St Vincent, and 6000 at Santa Maria, all in 1902; 1332 at Taal in 1911, 1300 at Merapi in 1930, 2942 at Mount Lamington in 1951, 1900 at Agung in 1963, and over 2000 at El Chichón in 1982. Plinian eruptions are probably the most destructive to property over the widest areas, their main threats springing from ashfalls, nuées ardentes and toxic gas, often spreading more than 30km from the vent. But prompt evacuation would often be possible, if the premonitory signs were correctly interpreted, even in the pitch darkness that these events usually produce. In this respect, the orderly, progressive evacuation of more than 100000 people from the danger zone around Pinatubo before the climax of the eruption in June 1991 to some 266 places of refuge was a triumph of planning as well as forecasting. Evacuation, however, would sometimes present enormous problems. About a million people live in the danger area if Vesuvius were to repeat its great exploit of AD79. Large eruptions near the sea often form tsunamis, which are propagated with little warning and can develop speeds of 250km an hour, so that any coastal zone facing a violent maritime volcano, such as Krakatau, is a potential target.

Many eruptions are dominated by the constraints of the relief around and upon them. Lava-flows almost invariably follow the valleys radiating from a volcano, but so, too, do lahars and nuées ardentes with more devastating effects. Large quantities of ice or water on the summit of an active volcano are a portent of great danger. The waters from the crater of Kelut in Java formed a lahar that killed 5000 people in 1919. Since its eruption in 1982, Galunggung has developed a summit crater lake which is already over 70m deep. A renewed outburst could generate a lahar which valleys would direct to Tasikmalaya, a town of 100000 inhabitants lying only 20km away. A volcanic danger can also be far away. The lahars radiating from Nevado del Ruiz claimed most of their 25000 victims some 60km from their source. Thus, any

240

Figure 15.1   House overwhelmed by lavas at Fornazzo, on Etna in 1971.

valley that is soon likely to receive a lahar or nuée ardente is a very dangerous place in which to linger.

On the other hand, mountains or ridges can protect areas and divert danger. The southernmost parts of Guadeloupe are sheltered from the Soufrière by the Monts Caraïbes, and the ridge of Monte Somma forms a protective arm around the northern flanks of Vesuvius. Huge volcanoes, such as Hawaii or Etna, are so large that the chances of any one locality being soon overrun by a lava-flow are small. It has been calculated that only half the unpopulated central area of Etna above 2000m will be covered by lava in the next 250 years, and it will take lava-flows 2000 years to cover half the populated region below 1000m. The risk of death is even smaller because these eruptions are predominantly effusive. Flank valleys, especially on the southeast and at the exit to the Valle del Bove, the great eastward-facing depression on Etna, are the most likely to be invaded by lava-flows, as the threats and damage to Mascali, Milo, Fornazzo and Zafferana have demonstrated during the present century (Fig. 15.1). But evacuation can be effected in time, always provided that the people are willing to leave. Paradoxically, Etna represented a greater danger to tourists visiting for the day than to those who spend their whole lives in villages on its flanks. Tourists used to visit the summit, far from all settlements, where the Voragine and the Bocca Nuova craters have been the scene of several sudden phreatic explosions. It was only after nine tourists were killed on 12 September 1979 that the extent of this threat really came to light and access to the summit was restricted.

Hazard assessment maps

Once a volcano has been studied, hazard assessment maps may be produced defining the areas of greatest danger and highest risk, although much prediction still depends on good judgement and qualitative evaluations by experts (Fig. 6.25). The great value of hazard assessment maps is that they have immediate impact by presenting a threat in a clear-cut way that the layman can appreciate. They cannot, of course, indicate when an eruption will happen, but can be invaluable public policy documents when one is deemed imminent. But, however colourful the map may be, a danger expressed in geological terms in a scientific periodical will not be perceived as much of a hazard at all. To be useful, a prediction must be clear and very precise, dated preferably in terms of days rather than weeks, and must identify the danger zones exactly. Otherwise, impact on the population and planners will be lost, and the forecast cannot be acted upon. Such inaction will not, of course, stop the volcano from erupting – whereupon governments and Earth scientists may then be criticized for not keeping the people informed. A map and televised film of volcanic hazards were major factors in demonstrating to people and authorities alike that Pinatubo presented a grave danger to the local inhabitants. The ensuing well organized evacuation must have saved thousands of lives.

Surveillance of volcanoes

What are believed to be the more dangerous volcanoes can be continually monitored by accurate measurements. Many surveillance methods are expensive and require skilled operators with sophisticated instruments in observatories. It is thus not really surprising that the United States, Russia, Iceland, Italy and Japan have been in the forefront of these investigations. The first of Japan's seven volcano observatories and 15 observation stations was founded in 1911 at Asama. Kilauea in Hawaii has been in intensive care for over 40 years; the former Soviet Union kept a careful eye on the volcanoes in Kamchatka for a similar period; and a remarkably varied surveillance network was established around Pozzuoli, in Italy, after its two phases of uplift (Figs 15.2, 15.3). But many countries, often those with the most dangerous volcanoes, are so poor that monitoring volcanoes cannot yet rise high amongst their priorities. However, with hindsight, constant surveillance of Nevado del Ruiz and its companions would have been much cheaper than the loss of 25 000 lives and damage estimated at US$1000 million.

The point of surveillance is that the ascent of magma causes a range of measurable repercussions which become more frequent and more intense as the eruption becomes imminent. The magma causes continual and increasing earth-tremors, deforms the volcano, increases and changes fumaroles and sometimes heats the rocks. These signals can only usually be picked up effectively through constant surveillance, because many of these warnings only become apparent a few months, weeks or even hours before the eruption starts. Moreover, the warnings sometimes

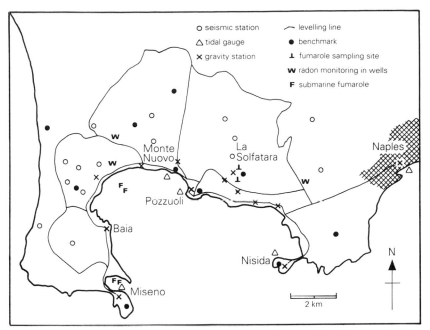

Figure 15.2 The varied surveillance network around Pozzuoli, Phlegraean Fields, Italy.

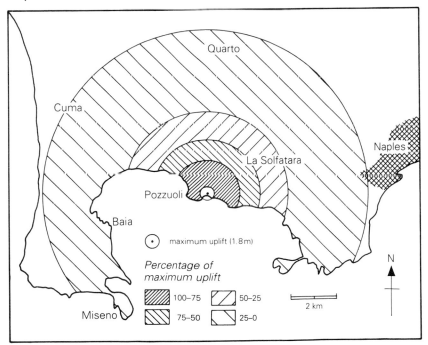

Figure 15.3 Uplift around Pozzuoli, 1982–4.

deceive: false alarms are more than twice as frequent as true eruption precursors in the West Indies, for instance. Surveillance, therefore, has to be not only continuous, but also sophisticated, if it is to form a sound basis for the short-term prediction of eruptions.

Seismographic monitoring

Shallow volcanic earthquakes are perhaps the most frequent and reliable sign that a volcano is about to burst back into life. Thus, amongst the continuous methods of surveillance, seismographic monitoring is a *sine qua non* of efficient short-term eruption prediction. In many cases it can be done by automatic telemeter recorders, relayed for immediate interpretation to an observatory. In Hawaii, for example, which has 51 such seismometers, data are also recorded on film and in digital form on magnetic tape. Rising magma causes moderate earth-tremors less than 5 km beneath the volcano that are clearly distinct from severe deep-focus earthquakes. The tremors often indicate that magma is approaching the surface when their frequency and intensity increase more than a hundred-fold (Fig. 6.26). Hawaii has such a fine network of seismometers that the epicentres of the tremors can be plotted with great accuracy. The epicentres often migrate along fissures ahead of the magma so that the path to the future point of eruption can be traced. Activity on Kilauea in January 1983 and on Krafla, in Iceland, between 1975 and 1984 was forecast in this way. Many small eruptions of the dome formed since 1980 at Mount St Helens have also been predicted to within a day, mainly when seismographic surveillance revealed that magma was rising.

Earth-tremors are sometimes accompanied by a kind of volcanic resonance with low-amplitude harmonic vibrations. These occurred before the eruptions at Klyuchevskoy in 1972, at Mount St Helens in 1980, at Nevado del Ruiz in 1985, and at Pinatubo in 1991. It is also believed that abnormally high earth tides, produced by lunar attraction on the crust, can precipitate eruptions of Augustine, in Alaska, provided the magma has already climbed high enough up the vent. Earth tides also influenced 14 of the 17 eruptions of the dome of Mount St Helens between 1980 and 1987.

Tiltmeters

As magma rises within a volcano, seismic activity is often accompanied by ground deformation that is measured by tiltmeters, which are very accurate levels, basically composed of three graduated pots arranged in a triangle and filled with water or mercury, which measure the tilting of the land with an accuracy of 1 mm in 1 km. Similar dry tiltmeters with levels and invar rods and electronic tiltmeters, which can record ground movements very rapidly, have also been installed on several volcanoes. They all measure tilting when rising magma makes a volcano bulge. This extraordinary process takes place slowly as the eruption looms. The summit of Unzen in Japan bulged out by 50 m in the three days before it emitted its initial

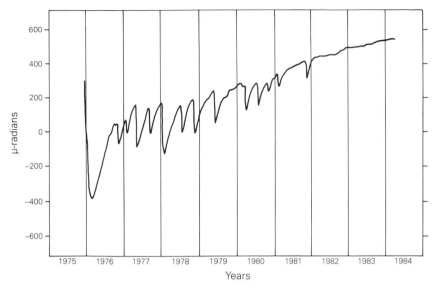

Figure 15.4 Inflation–deflation sequences at Krafla volcano, Iceland, 1975–84.

nuées ardentes on 24 May 1991. The much broader summits of both Kilauea and Mauna Loa swell by about 1 m before an eruption and subside again afterwards. At Krafla, tiltmeters were linked to triangulation surveys at Lake Myvatn (Fig. 15.4). The incursion of magma into the shallow reservoir slowly inflated the volcano. More rapid subsidence then deflated it as the magma migrated sideways along fissures and often emerged more than 10 km away. Thus, rapid deflation showed that Krafla was about to erupt on its flanks. The most spectacular instance of bulging – and with more disastrous results – occurred on the much steeper slopes of Mount St Helens in the two months before the 1980 eruption. The bulge could be seen by the naked eye and was constantly measured by laser. However, when the eruption was triggered by the earthquake and landslide, the rate of swelling was not itself increasing. The measurement of the bulge was not, therefore, telling the whole story, which again illustrates the need for various forms of monitoring on each volcano. In Hawaii and in Iceland, for example, seismometers and tiltmeters are used in conjunction with repeated geodetic levelling by lasers.

Some marked episodes of ground deformation and seismic activity have not been followed by the expected eruption. The most notorious instance in recent years was centred on Pozzuoli, in the Phlegraean Fields west of Naples, in 1969–72 and in 1982–4 (Fig. 15.3). The maximum uplift in the first episode amounted to 1.7 m and that of the second to 1.8 m. Ground movements related to magma displacements in the area have often been registered by changing water levels for 2000 years in the old Roman market known as the Temple of Serapis in Pozzuoli. One displacement in the 16th century culminated in the formation of Monte Nuovo in 1538. Thus, there seemed every reason for the authorities to act when the deformations began,

245

for 400000 people lived in the mobile zone. Pozzuoli was twice evacuated, but polemic succeeded panic when no eruptions ensued.

Gas and steam emissions

Increased emissions of gas and steam from fumaroles, mudpots and solfataras have long been known to herald eruptions because they often show that the magma is rising closer to the surface. Such information is often difficult to collect, however, because the emissions corrode instruments and are noxious to their observers, although these problems can be overcome to some degree by using infrared analysis and electrochemical sensors set at safe distances. Increases in carbon dioxide, sulphur dioxide or hydrogen sulphide emissions, or higher water temperatures and gas pressures are particularly significant. But constant monitoring of all the fumaroles in the world would be a vast and expensive programme, where diminishing returns would soon set in. Fumarole monitoring thus has to be selective to be worthwhile. Greater fumarole activity gave effective warning of the Askja eruption of 1961 in Iceland, for example, whereas emissions in the summit crater of Nevado del Ruiz virtually changed its waters to sulphuric acid before the eruption in 1985. And sulphur fumes from Montagne Pelée tarnished the silver in St Pierre in 1902. But gas emissions do not indicate an inexorable eruptive timetable. Increased fumarole activity on Mount Baker in Washington State in 1975–7, for instance, raised fears of an eruption that has not materialized.

Continual surveillance

Continual surveillance of individual volcanoes, then, often reveals that an eruption is likely, but not certain, to occur. But less than 50 out of 89 volcanoes designated as "high risk" were subject to detailed surveillance in 1983, for example, and it is, in fact, not even easy to identify the world's high-risk volcanoes. The two vigorous eruptions of 1991 illustrate contrasting facets of the problem. Pinatubo, without recorded activity for over 400 years, was not considered to be worth constant surveillance, in a country where Taal and Mayon had both erupted destructively during the present century. When Pinatubo threatened, quick installation of monitoring equipment fortunately made up for lost time. Although it had not erupted since 1792, Unzen, in Kyushu, was selected for surveillance in 1966 as one of 16 potentially dangerous volcanoes amongst the 83 active in Japan, because 14524 people had died in 1792 and the town of Shimbara now lies at its feet (Fig. 15.5). The rewards of continual surveillance were appreciated when it erupted in May 1991. Nevertheless, an unexpected nuée ardente killed three expert geologists and 38 other people on 3 June.

Figure 15.5   Unzen, Japan (by kind permission of Asia Air Survey Co. Ltd).

Surveillance by satellite

Remote-sensing imagery from satellites can sometimes show up thermal anomalies caused when rising magma heats the rocks close to the surface on active vents. In 1989, for instance, Landsat images of the central Andes revealed that there were, in fact, 60 active volcanoes in the region, rather than the 16 designated in the Catalogue of active volcanoes, published in 1963. Although satellite surveillance is costly and still in its infancy, it offers some of the best future prospects for forecasting eruptions. The Global Positioning System is also being used increasingly to monitor lateral and vertical ground displacements, including those around volcanoes, which might pinpoint future activity.

Preventive measures in response to an eruption

Throughout history mankind has tried to propitiate the gods or demons believed to inhabit active volcanoes. The Native Americans in Central America, for example, used to throw virgins into the crater of Masaya to calm its outbursts. The practice, however, can have neither encouraged virginity, nor appeased the volcano, for its eruptions have continued unabated. Even when an eruption can be predicted, it cannot be prevented. It is possible, though, to limit the damaging effects of many types of activity and evacuate the threatened population. However, it is easier for some populations to move than others. The rich can pick up their cheque books and drive off, but poor peasants in poor countries have to stay with their fields because they have nothing else.

Much also depends on what the volcano is going to emit. Unless they are encased in a valley, lava-flows are the easiest formations to influence, partly because they move so slowly and partly because of the way they solidify. However, even lava-flows cannot be mastered: they can only be tamed like lions confined to their cages. The lower slopes of Etna provide the most propitious terrain in Europe for experiments in flow control, as Diego Pappalardo demonstrated in 1669 when he dug holes into the solidified sides of the flow advancing on Catania to divert it into a new direction. Unfortunately, however, it turned towards Paterno, whose inhabitants forced Pappalardo to abandon the scheme. As a result, until recently, lava-flow diversion has been forbidden and left in the hands of the Almighty in Sicily, although sometimes the veil of St Agatha was paraded before a flow to induce it to halt. Efforts at diversion, however, were made in Spring 1983, when explosives were used to break up one side of a flow that was threatening seven villages, and four ash and cinder dams were constructed to contain the lavas nearer their source. But it took much effort to move $750000m^3$ of material in 50 days at a cost of US$3 million. It may, in fact, have been cheaper to let the lava-flow proceed unhindered, but the experience gained was invaluable when Zafferana was threatened.

The eruption that began on 14 December 1991 presented a major challenge because it was the largest lava emission on Etna since 1669, and, given its initial discharge rates and the local relief, it was thought that the flow would overwhelm Zafferana, a town with 7000 inhabitants (Fig. 15.6). Considerable resources were marshalled to prevent this disaster. A barrier of ash and cinders, 234 m long and 21 m high – the highest yet built to control lavas – was thrown across the narrow exit of the Valle del Bove to impede the current and buy time to defend Zafferana. When the lavas spilled over this barrier on 7 April 1992, three further dams were built down valley, but, to avoid possible legal consequences, care was taken not to divert the flow onto property that would not otherwise have been affected. Continued copious effusions, however, had taken the snout to within 700 m of the town by May 1992. Attention had, therefore, to be focused on the point of emission, 2000 m high and 8 km from Zafferana, where the current might be more effectively stemmed. On four occasions, concrete and steel blocks were dropped from helicopters onto the flow to impede its progress. Each intervention gave about two weeks of respite before the snout resumed its advance towards Zafferana. The fifth and most successful intervention took place on 29 May 1992. A deep channel was dug branching at a small angle from the flow. At its head, the solidified walls of the flow were first thinned and then blasted open. Two-thirds of the molten current opted for the new channel, enabling its former course to be blocked with boulders. Deprived of its supply, the natural snout immediately halted, while the snout of the diverted lavas began to advance over the initial solidified flow, 6 km from Zafferana. The struggle to protect the town had thus gained time and distance. All this effort had its reward on 1 June 1992, when the rate of lava discharge from the vent fell by

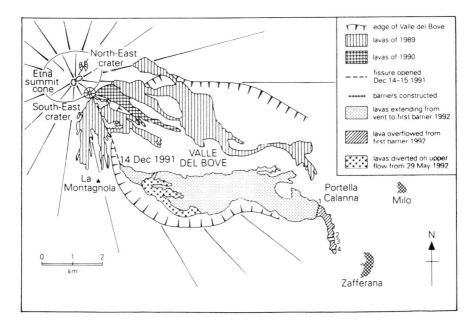

Figure 15.6 The eruptions of Etna (1989–92) and some results of efforts to control the flows in 1992.

half and further reduced the threat to Zafferana, before the eruption soon waned considerably and ceased altogether in March 1993. These were the most comprehensive lava control measures yet undertaken. The diversion near the vent took seven weeks of co-operative effort and could only have been undertaken in a sparsely populated area – and even then some complained that ecologically valuable habitats had been destroyed . . .

In spite of Etna's marked increase in activity since 1971, the people living on it usually have a respectful and fatalistic attitude to the volcano. "Etna is like a mother to us, she feeds and protects us, even though she is now acting like a cruel stepmother", said an inhabitant of Zafferana as the lavas advanced. Indeed, evacuation of homes has never seemed a serious option until lava has entered the kitchen; and the owner of one of the few houses to be overwhelmed by the flow in 1992 left an offering of bread and red wine on the table – and an imprecation against the government written on the wall.

The measures that can be taken against docile lava-flows would have little effect on lahars, jökulhlaups, blasts, poisonous gas, or nuées ardentes. Settlements in valleys radiating from dangerous strato-volcanoes would have to be surrounded by huge fortifications like the grandest of medieval citadels – and they would probably be thrown down or overridden anyway. After all, the nuée ardente from Montagne Pelée on 8 May 1902 had no difficulty in demolishing every building to its very foundations in the northern part of St Pierre. The only palliative to these lethal and fast-moving features is instant departure from the danger zone.

Falling ash causes damage to property but relatively few deaths. When ash rains down on a settlement, the inhabitants often panic and flee, especially since it can bring suffocating darkness too. Fear may also save lives, because the weight of ash accumulated often causes rooftops to collapse and bury those who stay behind. Fine ash spreads as all-penetrating clouds, choking humans, their animals, their cars, and even, on occasions, the engines of their jet aircraft. Few effective measures, except perhaps shovelling, can be taken against ash, as both the Pompeiians and the people of east Washington State have discovered in their time. The main threat from ash comes when it destroys crops and animals, such as when ash from Irazu destroyed the Costa Rican coffee crop in 1965. In the past, ashfalls have provoked widespread famines, most notoriously when the eruption of Tambora in 1815 caused 92000 deaths in Indonesia. Replacement of destroyed crops and animals, resettlement and international aid are usually the sole solutions. As a consolation for the hungry, however, ash increases future yields by adding valuable nutrients to the soil. Indeed, it is reported that some Papuan tribes pray for ashfalls to make their crops flourish.

The most sophisticated methods of disaster alleviation or population protection have been developed in Japan, where Sakurajima, in particular, has virtually become a surveillance and disaster-plan laboratory that could serve as a model for the rest of the world. Sakurajima has been continually active since 1955 and its slightest whim can be forecast and analyzed a few hours in advance in the world's most modern volcano observatory. The government has spent a fortune on counteracting the effects of these eruptions, but does not apparently offer financial help to those who wish to leave the area and cannot sell their homes. Closely spaced concrete volcanic bomb shelters have been built along the roads; children wear helmets to go to school; many expensive dams and canals to slow down and divert lahars have been constructed; all information is computerized, televised and automatically transmitted to the competent authorities. Every year a full-scale evacuation is rehearsed so that everyone knows where to go and what to do when the alert is given. Temporary lodgings are always ready. Few populations are so well protected . . .

Prediction and society

If predicting an eruption poses difficult volcanological problems, communicating predictions to the threatened society introduces a whole range of entirely different social, political and psychological questions. The threatened population wants simple answers to some apparently simple questions. When will the eruption start? What will emerge from the volcano? Where will the materials go? What will they do? Will we have to leave home? How long will the emergency last? Who will protect our property while we are away?

Such questions cannot be answered without caveats and provisos that would at best confuse, and at worst, alienate the questioner. Moreover, the Insiders (the Earth scientists) often write in learned journals and speak in a jargon which is incomprehensible to the society at risk. This society is composed of Outsiders, who need clear advice about what to do, and who largely depend on the media for their information. Apart, possibly, from in our work or hobbies, we are all Outsiders most of the time. But Outsiders are not stupid: they simply do not have the information upon which to act rationally. Outsiders, however, only respond to threats that they perceive, and they commonly suffer from disaster myopia – like drivers exceeding the speed limit in fog. For instance, the repose of Vesuvius since 1944 seems to have obliterated any recollection of past eruptions from the minds of both the authorities and the people living on its flanks, and towns are now rapidly expanding there as if the hazard had vanished. If the people living at the foot of Vesuvius cannot perceive a threat from a volcano, then who will?

Earth scientists must clearly communicate their predictions to local administrators and politicians and try to ensure that they act upon them. Unfortunately, they share neither the same perspectives, nor understanding of the subtleties of each other's problems. The authorities dislike the panic and disorders that a large-scale evacuation of population would almost inevitably entail; they also fully realize that they will be blamed for any miscalculations. Earth scientists, on the other hand, very rarely suffer the wrath of the voters. As the eruption of Montagne Pelée reached its climax in the electoral period in 1902, the opposition urged evacuation, but the governor of Martinique came to St Pierre to demonstrate his solidarity with the citizens and his confidence in their future. He died with them on 8 May. The accounts of the eruptions of Lanzarote, Heimaey, Nevado del Ruiz and Pinatubo illustrate the varied possible administrative responses to major crises.

251

Evacuation of a large population may be necessary in the most dangerous outbursts. The problem is particularly acute where lahars or nuées ardentes are likely to
occur, because they are often most severe in the initial stages of a magmatic eruption
and can sometimes give little warning. Governments should, therefore, make plans
in advance, train the population in evacuation drills, set up refuges and move people
to them, perhaps at very short notice, as soon as the eruption threatens. Quite apart
from other considerations, this may be tantamount to maintaining a spare town!
The experts will probably be unable to forecast the date of an eruption within a few
days, and the major administrative problem in this context is probably to decide
when such a displacement is worthwhile. There was a notorious case in Guadeloupe
in the West Indies in 1976, when two French teams of geologists disagreed about
the potential dangers that the eruption of the Soufrière presented. One group, believing that a catastrophe was imminent, advocated evacuation of the nearby town
of St Claude. The other group averred that an evacuation was unnecessary, because
the worst was already over after the initial phreatic eruption on 8 July 1976. Close
study of the ash expelled, and of the geological record, where nuées ardentes were
rare, indicated strongly that St Claude would not suffer the same fate as St Pierre.
The first group proved more persuasive, however, and the population of St Claude
was evacuated for several months. When the feared nuées ardentes did not emerge,
and the second group of scientists was proved right, considerable polemic raged
amongst the population, especially after their original panic turned to dissatisfaction.
Litigation eventually took place between the leaders of the two groups. It is interesting to speculate about how the remaining inhabitants of Pozzuoli will greet the
next episode of ground deformation after the last two false alarms. Clearly, the psychological barriers surrounding danger perception will always be raised whenever a
predicted eruption fails to materialize. Subsequent forecasts may not be believed,
and the threatened people may well then feel more confidence in pundits or sorcery.

The forecasting problem is not finished, however, when the population has been
evacuated. The end of the danger must also be forecast, so that the people can safely
return home. Unfortunately this is one of the more difficult tasks in the repertoire
of prediction. Governments are naturally unwilling to assume responsibility of this
kind, but equally naturally wish to be seen to be always doing their utmost. The
people's gratitude for being saved from a dangerous situation soon turns to unrest if
the emergency "lasts too long". The authorities also have the additional problem of
feeding, housing and clothing the evacuees, who may, in fact, only accept the
inconvenience of evacuation with docility for a short while. Not all the evacuees
will be offered homes overlooking Central Park . . . . The political and social needs
of evacuees stretch the limits attained by eruption forecasting and raise manifold
problems that are far beyond the expertise of Earth scientists.

Three quarters of the world's dangerous volcanoes are situated in poor countries
which may prefer to spend limited resources on famine relief, disease prevention, or
flood hazards rather than on monitoring volcanoes that may not be perceived as
dangerous in the short term. Even relatively wealthy Italy only set up a Ministry of
Civil Protection and a National Volcanology Group in 1981. Before El Chichón

252

erupted, Mexico, with over a dozen active volcanoes, had no co-ordination of civil defence and scientific experts, and no contingency plans for evacuation, shelter or resettlement. Similar deficiencies prevailed in Colombia in 1985, where they were compounded by administrative imbroglio and conflicting advice from both experts and non-experts. The fate of Armero shows what a failure of danger appreciation and disaster planning can entail. On the other hand, Indonesia and the Philippines have paid greater attention to eruptions, so that many lives were saved around Galunggung in 1982 and Pinatubo in 1991 through successful communication between experts and the civil and military authorities. Pinatubo erupted far more vigorously than Nevado del Ruiz, and in a far more densely populated area, but the Colombian volcano caused 80 times as many fatalities. The disaster averted at Pinatubo perhaps represents the first major victory for preventive action in the face of volcanic eruptions. Perhaps the lessons to be learned from the avoidable tragedy at Armero and from the tragedy thwarted at Pinatubo will ensure that the eruption of Nevado del Ruiz will be the last catastrophe wrought by lack of information or communication in the face of a volcanic onslaught.

It was easier at Mount St Helens because the numbers involved were small, communications were better and the danger was more obviously understood, in spite of extraordinary statements in the media. In April 1980, public access to the mountain was restricted, property owners were asked to leave, and most did so, albeit reluctantly. Only a few hundred people were involved, and most had other homes and their own transport. They were also supported by many efficient organizations and a wealthy state. When it became clear that an eruption was imminent, the Governor of Washington State arranged for police-escorted convoys of owners and youth-camp managers to pick up their possessions. Nevertheless, a sheriff rightly remarked that they were playing Russian roulette with the volcano. Their first trip took place on Saturday 17 May and a second was planned for 10.00 on Sunday 18 May. As the great eruption took place at 08.30, 200 people won their game of Russian roulette by an hour and a half.

As a final note of caution, it is important to remember that not all recent destructive outbursts have been predicted. An apparently extinct volcano has erupted on average every five years even during the present century. Mount Lamington in Papua–New Guinea was not even known to be a volcano when it erupted in 1951, and Bezymianny, scorned amongst its Kamchatkan fellows with the designation meaning "no-name", burst into life to produce one of the greatest explosions of the century in 1956. More than a dozen of the violent eruptions of this century have come from volcanoes that had apparently shown no signs of activity during historic time. Pinatubo joined this impressively disconcerting line in 1991. Predictions can also be inaccurate. The 1980 eruption of Mount St Helens was forecast, but its intensity and nature was not: no-one expected a climax without parallel in its 30 000 year history. Predictions about Montagne Pelée in 1902 were wholly wrong, but the forecasters scarcely had a viable methodological basis on which to make them. The death of the geologist David Johnston, backed by all the weight of American technology at his observation post 10 km from Mount St Helens on 18 May 1980, and the death of the vastly experienced Maurice Krafft on Unzen on 3 June 1991 show that much progress still has to be made in forecasting volcanic eruptions and the dangers that they represent. But the Earth sciences have come a long way since Pliny's sister first pointed out that odd cloud rising across the Bay of Naples in AD 79, and predictions no longer remain in the lap of the gods where they started with Hephaistos and Vulcan.

Glossary

Many terms are described more fully in the text.

aa A Hawaiian word describing lava-flows with a very rough, broken, angular surface.

andesite A common greyish volcanic rock, intermediate in composition between basalt and rhyolite and containing about 60 per cent by weight of silica. Emitted at temperatures of about 1000°C, it occurs in lava-flows, domes and fragments and is a common constituent of strato-volcanoes in subduction zones. It is fairly viscous and often forms aa lava-flows. Its fragments commonly result from violent explosions.

ash Pulverized volcanic rock exploded violently from a vent in fragments less than 2mm in size, often composed of rhyolite, dacite, trachyte or andesite. It can often form widespread blankets on land, and the finest particles may remain in the stratosphere for years.

ashflow A turbulent mixture of volcanic rock fragments and gas, expelled at high temperatures and great speed from a vent, which flows down slope and covers vast areas. Also called pyroclastic flow.

basalt A dark, pasty-grey volcanic rock, poor in silica (about 50% or less by weight) and relatively rich in iron, calcium and magnesium. By far the most common volcanic rock, forming the bulk of the ocean floors, and on land it occurs in many lava-flows, cinder cones, shield volcanoes and volcanic plateaux. It is usually hot (1200°C) and fluid when emitted, and flows 10km long are common. It is chiefly erupted in effusive emissions without violent explosions, from fissures as well as single vents.

block A large angular lava fragment, often 1m or more across, expelled during an eruption.

bomb A large, rounded lava fragment, often 1m across, expelled during an eruption, which develops a rounded or almond shape when twirled through the air. Some bombs have a characteristic breadcrust surface, others resemble cauliflowers or cow-pats, depending on the way in which they solidify.

caldera A large almost circular or horse-shoe shaped hollow, several kilometres across, formed mainly by collapse into the magma chamber, but also by great volcanic explosions. Usually bounded by steep enclosing walls and formed most often on strato-volcanoes. The term is derived from the Spanish word for cauldron. The Portuguese spelling, "caldeira", is sometimes used.

cinders Fragments of lava, between 64mm and 30cm in size, that are expelled by moderate explosions, which often form cones. They are usually light and riddled with gas-holes. They are most often composed of basalt or andesite and are frequently interbedded with lapilli. They are also commonly known by their Italian name of scoria.

cinder cone A steep conical hill, usually less than 250m high, with straight slopes at the angle of rest of the loose materials composing the cone. Formed above a vent when moderate repeated explosions accumulate layers of cinders, lapilli and ash.

crater A bowl-shaped hollow in the summit of a volcano which lies directly above the vent from which fragments are ejected.

crust The solid outer layers of the Earth forming both the continents and the ocean floors.

dacite A pale volcanic rock, rich in silica (60–65% by weight), which is viscous and slow-moving with a temperature of usually about 800–900°C when emitted. A common constituent of domes, it is often also involved in violently explosive eruptions.

dome   A rounded, convex-sided, dome-shaped mass of volcanic rock which is usually silicic and too viscous to flow far from the vent. Often formed on strato-volcanoes towards the end of an eruption. Frequently composed of dacite, phonolite, trachyte or rhyolite.

dyke   A vertical or steeply inclined sheet of magma injected along, and solidifying in, fractures in the Earth's crust. Dykes vary from a few centimetres to 10 m or more in thickness and can be many kilometres deep. Volcanoes are fed by deep dykes, and other dykes, in turn, branch or radiate from them.

eruption   The way in which gases, liquids and solids are expelled onto the Earth's surface by volcanic action ranging from violently explosive outbursts to noiseless, effusive outflow.

effusion   An eruption of lava which takes place with few or no gaseous explosions and thus most commonly gives rise to lava-flows.

fissure   A deep crack, fault, fracture, or cluster of joints, cutting deep into the Earth's crust, up which magma may rise. A fissure usually gives rise to effusive emissions, which may be accompanied by more explosive eruptions forming cones of cinders or spatter.

fragments   Ash, bombs, cinders, lapilli or pumice shattered by explosions during an eruption. They are the main constituents of cinder cones and many strato-volcanoes. Also called pyroclasts and tephra.

fumarole   A vent giving off gases or steam and often surrounded by fragile precipitated crystals. Occurs on active and dormant volcanoes.

geological time   The whole history of the Earth, extending back about 4600 million years, which is revealed by the study of the rocks of the Earth's crust.

granite   A coarse-grained crystalline rock composed chiefly of quartz and feldspars. It forms when rhyolitic magma solidifies and crystallizes below the Earth's surface, and is only exposed by erosion.

historical time   The timespan during which events have been recorded, in however fragmentary a fashion, by observers. In the Mediterranean World it may reach back 3000 years, whereas in the New World it may be less than 200 years.

hornito   A small cone or mound of spatter, usually less than 10 m, but occasionally 100 m in height. They are notably rough and steep-sided, whatever their size. The name is derived from the Spanish word for little oven, which smaller hornitos resemble.

hotspot   A stationary plume of convectively rising mantle which generates chains of volcanoes as the plates move over it.

island arc   A gently curving chain of volcanic islands rising above sea level from the ocean floor, formed when an oceanic plate is subducted beneath another. Volcanic chains are their equivalent on land.

lahar   An Indonesian word used to describe a volcanic mudflow commonly formed when an eruption melts part of an ice-cap, or disturbs a crater lake. They travel down valley at high speed and often cause much damage. Similar features, formed beneath Icelandic ice-caps, and probably containing a greater proportion of meltwater, are called jökulhlaups.

lapilli   Small pea- or nut-sized volcanic fragments between 2 mm and 64 mm across, expelled in a molten or nearly solid hot state during moderate eruptions and accumulating in layers within cones. The term derives from the Italian word for little stones.

lava   Molten rock or magma which reaches the surface and solidifies on cooling. Lava occurs as flows, domes, fragments within cones and as pillows formed on the ocean floors.

lava tube   A linear hollow formed when molten lava flows away beneath an already solidified outer crust. They are the chief way in which pahoehoe flows travel across the land surface. Larger forms are sometimes called lava tunnels.

lithosphere   The solid outer shell of the Earth which is broken into plates, which is composed of the Earth's crust and the solid outermost layer of the upper mantle.

maar   A German word used to describe an almost circular crater, often about 1 km across, enclosed by steep, inward-facing slopes, formed mainly by phreatic eruptions. They may or may not be bordered by a ring or crescent of fine fragments, sloping gently outwards from a low crest overlooking the crater. The crater is usually filled with rainwater, forming small lakes from which the name is derived.

magma   Hot, mobile rock material formed by partial melting of the mantle, commonly at depths between 70 km and 200 km. It is composed of hot, viscous, liquid material containing hot, but still solid, crystals, or rock fragments and small proportions of included gases. Under high pressures at depth, it behaves more as a plastic than a liquid. It is less dense than the materials surrounding it and is thus able to rise slowly towards the Earth's surface by buoyancy. If it overcomes the resistance of the rocks of the Earth's crust, it erupts in a fluid state, releasing its contained gases with varying degrees of explosivity and emits lava in flows or fragments. Magma which cools and solidifies before reaching the surface forms intrusive crystalline rocks such as granite, which are only exposed at the Earth's surface after the erosion of the overlying rocks.

magma reservoir   A large zone of ill defined fissures and cavities in the lithosphere where rising magma halts for varying lengths of time. Reservoirs are most often a few cubic kilometres in volume and situated usually between 2 km and 50 km in depth.

mantle   The hot, but not wholly mobile layer of the Earth situated below the Earth's crust and which envelops the Earth's core.

mid-ocean ridge   A ridge on the ocean floor where volcanic eruptions generate new oceanic crust and where two adjacent plates diverge.

nuée ardente   A French term used to describe an incandescent cloud or glowing avalanche of hot gas and fragments of all sizes, including ash, pumice and rock debris in an aerosol-like emulsion expelled by explosive eruptions, which travels across the ground at very high speeds and gives off glowing, billowing clouds. It is, perhaps, the most dangerous of all the forms of volcanic eruptions. Also called pyroclastic flow.

obsidian   Volcanic glass. A dense, shiny black or brown, glassy and rare form of rhyolite, which rises rapidly and cools rapidly, and is usually too viscous to flow far from the vent. It forms domes, mounds and short, rugged lava-flows.

pahoehoe   A Hawaiian word used to describe smooth, gently undulating, or ropy lava-flows often having a typical sheen. Formed most often by basalts emitted in a hot fluid state.

peridotite   A rock forming much of the mantle which is rich in olivine (peridot).

phreatic eruption   A sudden violent eruption chiefly of steam that emits newly shattered fragments of older solid rocks. Caused when rising new magma meets waters percolating

257

downwards. A phreatomagmatic eruption also emits fragments of new lava.

plate   Usually rigid slabs into which the lithosphere is divided. Their edges constantly diverge or converge and plunge beneath each other. All are composed of oceanic crust and some also carry continental crust. Between 10 and 15 major plates are generally recognized.

phonolite   A pale volcanic rock that is rich in sodium and potassium but poor in silica, which is usually emitted at temperatures about 1000°C. It is viscous and occurs in domes and in rugged lava-flows. It is so named because it often breaks into plates which give a characteristic ringing sound when struck.

pumice   Very pale volcanic fragments riddled with gas-holes, formed by the expansion of contained gases as the magma reaches the surface, and exploded very violently over vast areas during an eruption. Most pumice floats on water, and sometimes forms ephemeral floating islands after eruptions at sea. It varies from small fibrous chips to knobbly lumps and often resembles solidified foam. It is commonly composed of rhyolite, dacite, trachyte or phonolite.

rhyolite   A pale volcanic rock very rich in silica (over 70% by weight), and also rich in sodium and potassium, which is emitted at temperatures of about 700–800°C and commonly forms extensive pumice and ashflows when expelled as fragments, but it also forms viscous lavas forming domes and stubby flows.

seamount   A volcanic mountain found below sea level, and especially common in the Pacific Ocean. They often rise 2000 m from the ocean floor. Active seamounts are being built up at present by submarine eruptions and could eventually form new volcanic islands. They may also be extinct, submerged remains of old volcanoes.

shield   A large, gently sloping volcano composed mainly of fluid basaltic lava-flows emitted from clustered vents, with relatively few fragmented layers.

silica   The molecule formed of silicon and oxygen (SiO_2) that is a fundamental component of volcanic rocks, and the most important factor controlling the fluidity of magma. Other things being equal, the higher the silica content of a magma, the greater its viscosity.

solfatara   An Italian word used to describe the quiet emission of sulphurous gases from a fumarole.

spatter   Lava fragments of cinder size, often emitted as hot clots in lava-fountains. They are still molten when they return to the ground and thus flatten out and form cow-pats. Spatter often welds together to form steep-sided cones and ramparts as well as hornitos.

strato–volcano   A large, often steep-sided volcanic cone composed of stratified, bedded layers of lava fragments and flows as well as many other volcanic products. Also known as composite volcanoes or strato-cones.

subduction zone   Where two plates converge and one plunges beneath the other into the mantle. The subducted slab stimulates melting in the wedge of mantle above it and helps form volcanoes. The slab's plunging action also generates deep-seated violent earthquakes.

trachyte   A pale, greyish volcanic rock relatively rich in silica (65% by weight) and in sodium and potassium, which is usually emitted at temperatures of about 1000°C. It is viscous, and can be involved in violent explosions. It also forms rugged lava-flows and domes.

tsunami   A Japanese term used to describe huge, rapidly moving sea-waves generated by violent eruptions or earthquakes. They increase in size and speed as they reach shallow water and often cause much damage and death on nearby coasts.

tuff cone   A steep, squat, conical hill, usually less than 300 m high, composed of innumerable thin layers of fine fragments with a deep, wide crater, formed above a vent where explosions are generated by the interaction of rising magma and shallow water at the Earth's surface.

tuff ring   A broad, circular accumulation of fine fragments, often 1 km or more in diameter, surrounding a broad, shallow crater. Both the outer and craterward-facing slopes are relatively gentle compared with those of a cinder cone, and the crater is much wider. It has the approximate proportions of a doughnut or motor-car tyre.

vent   The usually vertical conduit or pipe up which volcanic material travels through the crust to the Earth's surface.

volcanic chain   A series of volcanoes, arranged in a curve or straight line, erupted on the continents as a result of subduction. They are the land equivalents of island arcs.

volcanic gas   Contained in small proportions within magma and commonly comprising, for example, steam, sulphur dioxide and carbon dioxide. As the magma closely approaches the Earth's surface, the gases are exsolved and can become a major factor in the violence of eruptions. Because several such gases are toxic, they can also be important contributors to volcanic death tolls.

volcano   A hill or mountain formed around and above a vent by accumulations of erupted materials, such as ash, pumice, cinders or lava-flows. The term refers both to the vent itself and to the often cone-shaped accumulation above it.

Further reading

Several books provide information on many aspects of volcanoes. Each has a different emphasis, but most have lists of references on more specific topics.

Bullard, F. M. 1984. *Volcanoes of the Earth*. Austin: University of Texas Press. (Perhaps the most interesting general text.)

Decker, R. W. & B. Decker 1989. *Volcanoes*. San Francisco: W. H. Freeman. (Short, concise and very well written.)

Decker, R. W. & B. Decker 1991. *Mountains of fire*. Cambridge: Cambridge University Press. (A very well written summary of volcanic activity.)

Francis, P. 1976. *Volcanoes*. Harmondsworth, England: Penguin. (Aimed at the general reader.)

Green, J. & N. M. Short (eds) 1971. *Volcanic landforms and surface features*. New York: Springer. (Valuable for black & white illustrations and comments on volcanic relief.)

Krafft, M. & K. Krafft 1980. *Volcanoes: Earth's awakening*. Maplewood, NJ: Hammond. (An introduction aimed at the general reader.)

The following are rather more specialized books written in a more technical language:

Cas, R. A. F. & J. V. Wright 1987. *Volcanic successions*. London: Allen & Unwin. (Well illustrated, detailed foundation geological work.)

Francis, P. 1993. *Volcanoes: a planetary perspective*. Oxford: Oxford University Press. (A broad-ranging survey, which concentrates on volcanic deposits, and is written in language suitable for specialists.)

Hardy, D. A. (ed.) 1990. *Thera and the Aegean world* III [3 volumes]. Third International Congress, Proceedings. Santorini, Greece, 3–9 September 1989. London: The Thera Foundation.

Kilburn, C. R. J. (ed.) 1993. *Active lavas*. London: UCL Press.

Macdonald, G. A. 1972. *Volcanoes*. Englewood Cliffs, NJ: Prentice Hall. (US orientation.)

Ollier, C. 1988. *Volcanoes*. Oxford: Basil Blackwell. (Australian orientation.)

Rittman, A. 1962. *Volcanoes and their activity*. New York: John Wiley. (European orientation and now rather old fashioned.)

Williams, H. & A. R. McBirney 1979. *Volcanology*. San Francisco: Freeman, Cooper. (Has a geological bias and is written in technical language.)

Lists of active volcanoes are found in:

Simkin, T. L., Siebert, L. McClelland, D. Bridge, C. Newhall, J. H. Latter 1981. *Volcanoes of the world*. Stroudsburg, PA: Dowden, Hutchinson & Ross and the Smithsonian Institution (Washington DC).

The volumes of the *Catalogue of the active volcanoes of the world*, published by the International Association of Volcanology, give details of eruptions, rock analyses and, sometimes, maps, but are not usually as informative as they might seem:

Georgalas, G. C. 1962. *Greece*.

Imbo, G. 1965. *Italy*.

260

Macdonald, G. A. 1955. *Hawaiian Islands.*
Mooser, F., H. Meyer-Abich, A. R. McBirney 1958. *Central America.*
Neumann Van Padang, M. 1951. *Indonesia.*
Robson, G. R. & J. F. Tomlin 1966. *West Indies.*

Individual volcanoes or volcanic areas are treated in a readable way in:

Chester, D. K., A. M. Duncan, J. E. Guest, C. R. J. Kilburn 1985. *Mount Etna: the anatomy of a volcano.* London: Chapman & Hall.
Harris, S. 1988. *Fire mountains of the West.* Missoula, Montana: Mountain Press.
Herbert, D. & F. Barnossi 1968. *Kilauea: case history of a volcano.* New York: Harper & Row.
Lipman, P. W. & D. R. Mullineaux (eds) 1981. *The 1980 eruptions of Mount St Helens.* USGS Professional Paper 1250.
Luce, J. V. 1969. *The end of Atlantis.* London: Thames & Hudson. (Deals with Santorini and Krakatau.)
Macdonald, G. A., A. T. Abbott, F. L. Peterson 1983. *Volcanoes in the sea.* Honolulu: University of Hawaii Press.
Perret, F. A. 1935. *The eruption of Mt Pelée 1929–1932.* Washington DC: Carnegie Institute.
Simkin, T. & R. S. Fiske 1983. *Krakatau 1883: the volcanic eruption and its effects.* Washington DC: Smithsonian Institution.
Thorarinsson, S. 1969. *Surtsey: the new island in the North Atlantic.* London: Cassell.
Villari, L. (ed.) 1980. *The Aeolian Islands.* Catania, Italy: Istituto Internazionale di Vulcanologia.
Wood, C. A. & J. Kienle (eds) 1990. *Volcanoes of North America (United States and Canada).* Cambridge: Cambridge University Press.

The relationships of volcanoes to the environment are well treated in:

Blong, R. J. 1984. *Volcanic hazards.* New York: Academic Press.
Chester, D. K. 1993. *Volcanoes and society.* London: Edward Arnold.
McGuire, W. J., C. R. J. Kilburn, J. B. Murray (eds) 1994. *Monitoring active volcanoes.* London: UCL Press.
Press, F. & R. Siever 1986. *Earth.* San Francisco: W. H. Freeman.
Sheets, P. D. & D. K. Grayson (eds) 1979. *Volcanic activity and human ecology.* New York: Academic Press.

Many interesting articles about volcanoes appear from time to time in *Scientific American* and *Geographical Magazine*, including:

Decker, R. W. & B. Decker 1981. The eruptions of Mount St Helens. *Scientific American* (March).
Francis, P. & S. Self 1983. The eruption of Krakatau. *Scientific American* (November).
Hekinian, R. 1984. Undersea volcanoes. *Scientific American* (June).

The chief journals for research are *Bulletin Volcanologique*, now entitled *Bulletin of Volcanology*, and the *Journal of Volcanology and Geothermal Research*. Both contain valuable articles based on the most recent research, but they are written in a very technical way. In addition, shorter, rather more accessible information appears in *Earthquakes and Volcanoes*. From time to time research articles dealing with volcanoes appear in the USGS *Bulletin* and the USGS Professional Papers.

Index of volcanoes
and other features

General index